# FEMTOPHYSICS

*A Short Course on Particle Physics*

*Related Pergamon Titles of Interest*

BOOKS

BITTENCOURT
Fundamentals of Plasma Physics

BOWLER
Lectures on Special Relativity

BOWLER
Lectures on Statistical Mechanics

*DURRANI
Solid State Nuclear Track Detection (Principles, Methods & Applications)

*KHALATNIKOV: LANDAU
The Physicist and the Man (Recollections of L. D. Landau)

SERRA *et al.*
Introduction to the Physics of Complex Systems (The mesoscopic approach to fluctuations, non linearity and self-organization)

JOURNALS

Annals of Nuclear Energy

Plasma Physics & Controlled Fusion

Progress in Particle & Nuclear Physics

Full details of all Pergamon publications/free specimen copy of any Pergamon journal available on request from your nearest Pergamon office

*Not available under the terms of the Pergamon textbook inspection service

# FEMTOPHYSICS

## *A Short Course on Particle Physics*

by

M. G. BOWLER

*Department of Nuclear Physics*
*Oxford University*

PERGAMON PRESS

Member of Maxwell Macmillan Pergamon Publishing Corporation

OXFORD · NEW YORK · BEIJING · FRANKFURT
SÃO PAULO · SYDNEY · TOKYO · TORONTO

| | |
|---|---|
| U.K. | Pergamon Press plc, Headington Hill Hall, Oxford OX3 0BW, England |
| U.S.A. | Pergamon Press, Inc., Maxwell House, Fairview Park, Elmsford, New York 10523, U.S.A. |
| PEOPLE'S REPUBLIC OF CHINA | Pergamon Press, Room 4037, Qianmen Hotel, Beijing, People's Republic of China |
| FEDERAL REPUBLIC OF GERMANY | Pergamon Press GmbH, Hammerweg 6, D-6242 Kronberg, Federal Republic of Germany |
| BRAZIL | Pergamon Editora Ltda, Rua Eça de Queiros, 346, CEP 04011, Paraiso, São Paulo, Brazil |
| AUSTRALIA | Pergamon Press Australia Pty Ltd., P.O. Box 544, Potts Point, N.S.W. 2011, Australia |
| JAPAN | Pergamon Press, 5th Floor, Matsuoka Central Building, 1-7-1 Nishishinjuku, Shinjuku-ku, Tokyo 160, Japan |
| CANADA | Pergamon Press Canada Ltd., Suite No. 271, 253 College Street, Toronto, Ontario, Canada M5T 1R5 |



First edition 1990

**Library of Congress Cataloguing in Publication Data**

Bowler, M. G.
Femtophysics: a short course on particle physics: by M. G. Bowler.
p. cm.
Includes bibliographical references.
1. Particles (Nuclear physics) I. Title
QC793.2.B68 1990 539.7'2—dc20 89–28833

**British Library Cataloguing in Publication Data**

Bowler, M. G. (Michael George)
Femtophysics.
1. Elementary particles
I. Title
539.7'21

ISBN 0–08–036943–X Hardcover
ISBN 0–08–036942–1 Flexicover

*Front cover illustration*

A u quark in a proton radiates a hard gluon and subsequently annihilates with an antiquark in an antiproton to form a massive boson $Z^\circ$. The $Z^\circ$ decays into an e+e− pair. The intermediate bosons of the weak interaction were first observed at CERN through such processes.

*Printed in Great Britain by BPCC Wheatons Ltd, Exeter*

# PREFACE

Atoms are structures of size $\sim 10^{-8}$cm, consisting of electrons bound to a positively charged nucleus. The physics of these composite structures became well understood only half a century ago, with the development of quantum mechanics. The crude technology of atoms is prehistoric in origin — fire, metallurgy and more recently chemistry. A precision atomic technology is emergent, with the development of techniques for manipulating and imaging single atoms.

The nucleus of the atom is again a structure, of size $\sim 10^{-13}$cm (1fm) $-10^{-12}$cm. The components are nucleons, protons and neutrons. Nuclear technology exists, but there is no control on a scale of $10^{-12}$cm and present day nuclear technology may be compared with atomic technology at the level of mastery of fire. The physics of nuclei was largely understood over thirty years ago. At a fundamental level, study of nuclear structure led to the discovery of fields of short range which involve the exchange of internal quantum numbers between the sources.

Just as nuclear physics emerged from the periodic table of the elements and the probing of the nucleus with beams of particles, the rich structure exhibited by the strongly interacting particles, hadrons, among which are to be found the constituents of the nucleus, suggested many years ago that these particles are also composite structures. The probing of hadrons with beams of electrons, muons and neutrinos has revealed the structure and identified the constituents. Hadrons are structures of size $\lesssim 10^{-13}$cm composed of either quark-antiquark pairs (mesons) or three quarks (baryons). The quarks have spin $\frac{1}{2}$, electric charge $+\frac{2}{3}$ and $-\frac{1}{3}$ (in units of the electron charge) and are bound into hadrons through the operation of eight vector fields, quantum chromodynamics. There is as yet no technology on the scale of particle physics, $< 10^{-13}$cm.

The particle physics of today is the physics of spin $\frac{1}{2}$ particles, leptons and quarks, interacting through vector and axial vector fields, the free field quanta of which have spin 1. These fermions and bosons are at present treated as elementary and should they have structure the scale must be $< 10^{-16}$cm. The theoretical structure known as the standard model embraces quantitatively all known aspects of particle physics, and over the last ten years has been tested with increasing precision and has emerged with distressing success. The fundamental physics that has emerged from the study of particles is the physics of self-interacting vector fields, covered by the principle of gauge invariance.

Very many books have been written on the subject of gauge field theory. The majority have been written by theoretical physicists at a level which makes them inaccessible to most undergraduates. There are relatively few books which cover the general field of particle physics at (advanced) undergraduate level.

In the four years 1986–1989 I lectured on particle physics to Oxford undergraduates in their final year who had chosen to take an advanced optional course in Nuclear and Particle Physics. The subject is my own and I took the opportunity to make a fresh study of the subject in the light of the advances of the last ten to fifteen years. I developed a new course and here it is. I have restricted myself to those aspects of the subjects which I regard as established and likely to stand, regardless of developments in the future. I am as confident of the existence of quarks and gluons as I am of the existence of electrons and photons, less confident that the vacuum is inhabited by a Higgs field and I have no confidence that any of the tentative developments to which I allude briefly under Final Remarks will survive. With the exception of those remarks, I have confined myself to topics which I believe I understand. I do not understand grand unified theories in any detail, nor supergravity and superstrings at all. However, a publisher's referee remarked that superstrings should at least appear in the index ... hence much of Final Remarks.

My concern throughout has been with what I conceive to be the underlying physics of the subject and I have dispensed with formal mathematics wherever possible. There are no mathematics more advanced than volume integration and matrix multiplication. I have explicitly constructed the Clebsch-Gordan coefficients for SU(2) and SU(3) and made no appeal to formal group theory. I believe the book will be accessible to the advanced and enthusiastic undergraduate and that it will also be useful to graduate students. The first six chapters introduce the subject at an elementary level — and could be used as the basis for a short course — the remaining chapters are much more technical. The reader who finds the first chapters trivial should press on.

The first chapter introduces the fundamental particles and their interactions. The experimental significance of large centre of mass energy is discussed. Massive scalar fields and the (field theoretic) reinterpretation of negative energy solutions are introduced. The strength of the strong interaction is considered and this is further illustrated in Ch.2, which opens with a calculation of scattering through a Yukawa potential acting once. We thus obtain the propagator for virtual meson exchange and the connection between real and virtual particles is discussed in terms of Fourier components and annihilation and creation operators. Charge exchange interactions are introduced and nuclear charge independence is used to calculate the isospin coupling coefficients for $1 \otimes \frac{1}{2}$.

The muon and the strange quark were the first identified members of the second generation of quarks and leptons. Both decay weakly and introduce Ch.3. Dimensional considerations which are essential in discussion of strange particle decays lead naturally to the violation of conservation of probability by the pointlike Fermi interaction and hence to the intermediate bosons $W^{\pm}$ and the correct way to construct a dimensionless coupling for the weak interactions.

Ch.4 is a gallop through the properties of the forest of hadrons, including some discussion of how (over many years) these properties have been established. The singlet, octet and decuplet structures uncovered gave rise to the elementary quark model, which is introduced and criticised. In Ch.5 it is shown how the concept of confined colour augments the quark model and gives rise to verifiable (and verified) predictions in addition to solving the problems raised at the end of Ch.4.

Ch.6 is concerned with the evidence for the existence of tiny grains of momentum and energy, carrying electro-weak charges, within hadrons. An elementary discussion of form factors, deep inelastic scattering and the Drell-Yan process is given. (The evidence that these partons are light quarks is deferred until Chs.9 and 10.)

In Ch.7 the Fermi Golden Rule is derived within time-dependent perturbation theory and its validity and use without this framework discussed. The time evolution of an explicitly decaying state is used to find the width. The Breit-Wigner formula is obtained by a simple extension. The relation between resonance formulae and boson propagators in the time-like region is examined. The chapter closes with the introduction of invariant matrix elements and phase space.

The essential evidence for the existence of light quarks of spin $\frac{1}{2}$ cannot be understood without the Dirac equation. In Ch.8 I have constructed an explicit representation — that of Weyl — in the simplest way that I know, by starting with the Dirac equation for zero mass. The eigenstates can be represented by two-component spinors and are easily constructed; the rotational properties of spin $\frac{1}{2}$ emerge through this construction. Vector (or axial vector) interactions couple left-handed particle to left-handed particle .... It is easy to stitch together two Weyl equations and then reintroduce mass. The result is the Weyl representation of the $(4 \times 4)$ Dirac matrices, admirably adapted for calculation when mass is small or may be neglected. (I found to my delight that many relatively complicated problems in nuclear $\beta$ decay are easily solved in this representation by a little matrix multiplication, even though the mass of the electron may not be neglected.) I have tried to distinguish carefully between the properties **helicity** and **handedness**: a left **helicity** particle contains a right **handed** piece, which is lost only in the limit $v \to c$. I think that lack of a clear distinction between these properties is responsible for much confusion when attempting to understand the $V - A$ structure of the weak interactions.

The material of Ch.8 is put to work in Ch.9 in order to calculate explicitly the matrix elements for fermion-fermion interactions through vector (or axial vector) fields. The coupling recipe, together with the rotational properties of spin $\frac{1}{2}$, yields the essential features of $e^+e^-$ annihilation to a fermion-antifermion pair and of fermion-fermion scattering, without even $2 \times 2$ matrix multiplication. I have attacked the content of deep inelastic scattering by supposing that the nucleon consists of light quarks obeying the Dirac equation with minimal electro-weak coupling and obtaining the differential cross sections which are known (experimentally) to apply. In Ch.10 it is shown how comparison of deep inelastic electron (and muon) scattering with deep inelastic neutrino scattering establishes that the partons which couple to the electro-weak interactions are quarks.

Chapter 11 reverts to hadron structure and isospin. It is concerned first with distinction between those aspects of the formalism which are essential and those which are only conventional. The decay of hadrons into other hadrons must take place by creation of quark antiquark pairs from energy stored in the colour field, and the physical origin of isospin invariance is to be found in the amplitudes for creation of $u\bar{u}$ being (almost) identical with the amplitudes for creation of $d\bar{d}$. A systematic implementation of this rule leads to explicit construction of the isospin coupling coefficients (SU(2) Clebsch-Gordan coefficients) for $0 \otimes 0$, $1 \otimes 0$, $1 \otimes 1$. The famous rule that $(1,0) \otimes (1,0)$ does not couple to $(1,0)$ emerges as an exact cancellation of amplitudes. It is easy to extend such construction to take account of charge conjugation and hence extract the almost forgotten quantity $G$ parity. Perhaps this is doing things the hard way, but after I had devised this treatment (for a graduate course which I gave some years ago) I felt that I actually understood isospin (and $G$ parity) for the first time.

Chapter 12 establishes the content of an SU(3) of colour. First, the effects of interaction between quark-quark, quark-antiquark with two colours, colour exchange forces and colour independence are studied. This is little more than a relabelling of isospin. Three colours are then introduced, with colour exchange forces and colour independence. The relations among the couplings required for colour independence are solved and for vector colour fields both quark-antiquark and $qqq$ ground states are colour singlets. With colour confined within a flux tube a mechanism exists for driving all coloured states to infinite mass. A three quark colour singlet is antisymmetric under permutation of colour: the spin-space configurations of the quarks are dictated. Hyperfine splitting by chromomagnetic interactions is worked out for the low lying mesons. In Ch.13 this is extended to the three quark systems and the baryon magnetic moments are calculated in terms of the quark structure. The chapter ends with a discussion of the mixing of quark antiquark pairs through annihilation into colourless configurations of the colour fields.

Chapter 14 harks back to Ch.3: it is concerned with the properties of the massive $\tau$ lepton and the quarks $b$ and $c$. The Bohr atoms of QCD, $c\bar{c}$ and $b\bar{b}$ systems, are discussed in elementary but quantitative terms. The lowest lying states consisting of a heavy quark and light antiquark are considered briefly.

Chapter 15 is concerned with the weak interactions of quarks. The curious pattern of weak interactions connecting quarks in different generations is studied, leading to the famous GIM mechanism for the suppression of weak processes which change hadron flavour but do not change hadron charge.

Chapter 16 contains a discussion of weak isospin and the entanglement of the electromagnetic and weak interactions. It is shown how the assumption that a singlet gauge boson (a proto-photon) and the neutral member of a triplet $W^{\pm\circ}$ are mixed leads to the identification of the couplings of the quarks and leptons to the neutral intermediate boson $Z^\circ$. The predictions are tested and they work. The mechanism whereby weak isospin is broken, the proto-photon and $Z^\circ$ mixed and $W^\pm$, $Z^\circ$ acquire mass is introduced in general terms. This is an introduction to the Higgs mechanism but the apparatus of gauge field theory is left alone. The structure of the Weinberg-Salam model is discussed, again in elementary terms. I have devised a diagrammatic representation of the interaction of gauge bosons with a robust vacuum screening current which enabled me to understand (for the first time) the content of more formal development of the theory, and which is easily adapted to obtain the predictions of the theoretical structure for more complicated Higgs sectors than that of the standard model.

These sixteen chapters were originally in the form of sixteen lectures, each lasting one hour. I had to maintain a cracking pace to cover only the original material, and in preparing this book I have added more. Even so, some topics are omitted entirely. I have devised Problems, which appear at the end of every chapter, and many of these are invitations to explore omitted topics and can be solved with the techniques developed in preceding chapters. Those problems which are particularly difficult or time consuming have been indicated with a star.

The reader will need to be acquainted with electromagnetic theory, relativity and (non-relativistic) quantum mechanics. Some prior knowledge of nuclear and particle physics would help.

The style of these extensively edited lectures is essentially oral: I have tried to eliminate the inevitable infelicities without ruining the pace and impact of what is intended to be a compact introduction to what is known today of particle physics — physics at a scale of less than $10^{-13}$cm. It is remarkable that ancient notions of the nature of space and time and the relativistic quantum theory of fields have — so far — survived unscathed to a level of $\lesssim 10^{-16}$cm, $10^{-3}$fm.

# CONTENTS

# 1. AN INTRODUCTION.

## 1.1 What is a particle?

The subject of this book is called variously elementary particle physics, particle physics and high energy physics. It is the physics pertaining to distances less than or of the order of $10^{-13}$cm and the denizens of this world are by definition particles. We distinguish particles and elementary particles: our working definition of a particle is anything smaller than $10^{-13}$cm. This unit is known as the fermi and is also one femtometer; $10^{-15}\text{m} = 10^{-13}\text{cm} = 1\text{fm}$.

The vast majority of particles—the strongly interacting particles known as hadrons—are not elementary but are bound states of quarks. The particles still regarded as elementary, or fundamental, comprise leptons, quarks and the quanta of the gauge fields which mediate interactions among them.

The leptons are spin $\frac{1}{2}$ fermions which interact via electromagnetism and the weak interactions. There are three with electric charge $q = -1$ (in units of the electron charge) and three neutrinos:

| | | | | | | |
|---|---|---|---|---|---|---|
| $q = -1$ | $e$ | 0.511 MeV | $\mu$ | 106 MeV | $\tau$ | 1.784 GeV |
| $q = \ \ 0$ | $\nu_e$ | <18 eV | $\nu_\mu$ | <0.25 MeV | $\nu_\tau$ | <35 MeV |

The masses are given in units of rest mass energy.

The quarks are spin $\frac{1}{2}$ fermions which interact via electromagnetism, the weak interactions, and the colour fields. Each may carry one of three colours, usually denoted $r$, $b$, $g$, and these colours act as sources and sinks for the colour field:

| | | | |
|---|---|---|---|
| $q = +\frac{2}{3}$ | $u \sim 1$ MeV | $c \sim 1.5$ GeV | $t > 40$ GeV |
| $-\frac{1}{3}$ | $d \sim 6$ MeV | $s \sim 200$ MeV | $b \sim 5$ GeV |

There is little doubt that the quarks exist only as constituents of hadrons.

The ordinary world is made up from the first of the three generations alone—the others are transients (with the possible exception of the neutrinos).

## 1.2 Fundamental interactions.

The photon ($\gamma$) couples to the electromagnetic current and is the quantum of the electromagnetic field. There are three gauge bosons mediating the weak interactions—$W^\pm, Z^\circ$. The charged bosons $W^\pm$ mediate nuclear $\beta$ decay and change leptonic charge but not leptonic generation. The interactions are represented by the diagrams:

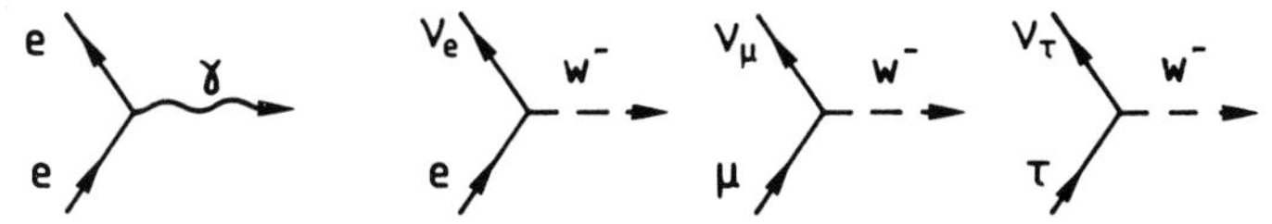


Fig.1.1

The coupling to leptons is universal and does not depend on the generation. Between the quarks, the charged current interaction predominantly operates within a given generation, for example

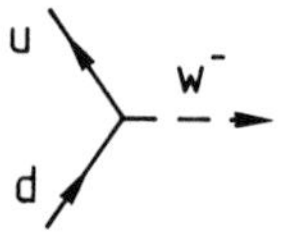


Fig.1.2

but the processes

(a) (b)

Fig.1.3

mix generations. The cross generation links attenuate as generation number increases: the rates for the processes illustrated in fig.1.3(a),(b) are suppressed relative to fig.1.2 by factors of 0.05, 0.005 respectively.

The photon has mass $m_\gamma < 10^{-15}$ eV and does not distinguish between left and right handed fermions†. The $W^\pm$ ($M_{W^\pm} \sim 80$ GeV) couple only to left handed fermions, thus violating parity to the maximal extent. The $Z^\circ$ couplings ($M_{Z^\circ} \sim 90$ GeV) are complicated, change neither charge nor flavour, involve only left handed neutrinos but both left and right handed massive fermions.

The colour couplings among the quarks seem to be universal, independent of flavour, and there are eight gauge bosons, for example

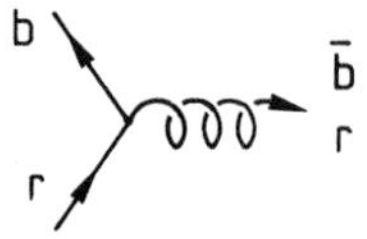


[$3 \otimes \bar{3} = 8 \oplus 1$. Experimentally there is no 1]

### 1.3 Confinement

Colour seems to be absolutely conserved and confined within a region of space of dimensions ~ 1fm.

The hadrons are colourless composites with (minimal) composition $q\bar{q}$ and $qqq$. These composites are the so called strongly interacting particles. The ($q\bar{q}$) set are bosons (integral spin) and strongly interacting bosons are called mesons; the ($qqq$) set are fermions (half integral spin) and strongly interacting fermions are called baryons. Note that with the interactions already discussed and no others, there must be at least one stable baryon; baryon number is conserved.

The composites have excited states and these are the particles which appear as resonances in hadron-hadron interactions. You (apparently) never get

† In the high energy limit a left handed fermion has spin oriented opposite to its direction of motion, a right handed fermion has spin parallel to its direction of motion.

quarks out on their own because (probably) the colour field, very like QED at short distances, squeezes itself into a tube of colour flux with constant energy/unit length. In the rapid stretching of such a string, colourless $q\bar{q}$ pairs are produced thereby snipping the string into little pieces—more hadrons... The energy density in the colour string is $\sim$ 0.9 GeV/fm—14 tonnes weight. This colour string is a useful concept unifying much of hadron physics.

### 1.4 Why is particle physics high energy physics?

We are very limited in the tools available for exploring this world. We need high energies for two reasons:

(1) To make objects of high mass (such as $W^{\pm}$, $t\ldots$)
(2) To resolve structure at small distances.

In both cases it is the centre of mass energy which counts. This is clear enough so far as making massive objects is concerned, but less clear as a condition affecting spatial resolution.

In optical microscopy we require $\lambda \lesssim d$ (the scale of the structure to be resolved) but there is a further condition. In order to resolve structure the objective must accept not only the forward diffraction peak from an illuminated object, but at least the first order in addition (the Abbé criterion). Thus a microscope must subtend an angle $\theta \gtrsim \lambda/d$; $\theta d/\lambda \gtrsim 1$. For particles, $\lambda = \frac{h}{p}$ so the condition is $p\frac{\theta}{h}d \gtrsim 1$, where $p\theta$ is $\sim$ the momentum transferred in the scattering. We can do no more than turn our probe around in the centre of mass

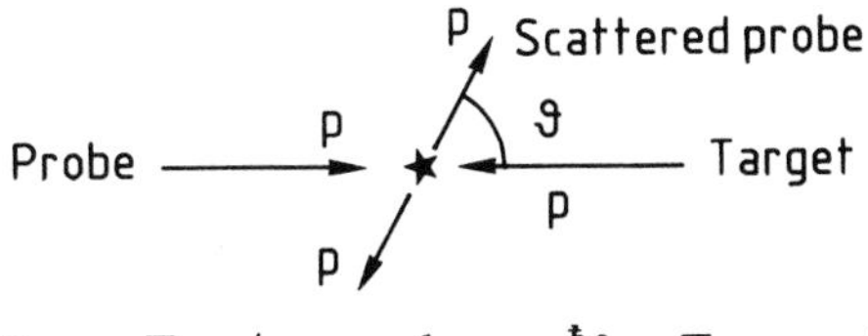


so that $(\Delta p)_{\max} = 2p \simeq E_{cm}/c$ : scale $\approx \frac{\hbar c}{E_{cm}}$. For resolution of structure at a level of 1fm, the momentum transfer must exceed $\sim$ 0.2 GeV/c and the centre of mass energy must exceed $\sim$ 0.2 GeV. This is the condition for femtoscopy. If the scale is $\sim 10^{-3}$fm then $E_{cm} \gtrsim 200$ GeV.

The prime movers are electron and proton accelerators, because electrons and protons are stable and exist in profusion. The highest energy proton beam is $\sim$800 GeV (Fermilab) and the highest energy electron beams are $\sim$50 GeV. From proton beams on fixed targets we derive secondary beams of various hadrons, of muons and neutrinos. The only particle fixed targets are protons, either free or bound in nuclei, neutrons in nuclei and electrons.

It is easy to calculate the centre of mass energy in terms of the beam momentum and the mass of the target. The quantity $(\Sigma E)^2 - (\Sigma \mathbf{p})^2$ is the square of a four vector and so invariant. In the centre of mass frame $\Sigma\mathbf{p}$ is zero and so we have, neglecting the mass of the projectile (and working in units with $c = 1$):

$$E_{cm}^2 = (\Sigma E)^2 - (\Sigma \mathbf{p})^2 \simeq (M_T + p)^2 - p^2 \simeq 2M_T p$$

| Beam energy (laboratory) | Target (at rest) | Centre of mass energy |
|---|---|---|
| 800 GeV | nucleon | 40 GeV |
| 800 GeV | electron | 1 GeV |
| 50 GeV | nucleon | 10 GeV |
| 50 GeV | electron | 0.2 GeV |

It is obvious why we also use colliding beam machines. Useful colliding beams of $p$ on $p$, $p$ on $\bar{p}$, $e^+$ on $e^-$ have been deployed as a result of superb engineering at SLAC, CERN, DESY, Fermilab... At Fermilab 800 GeV $p$ on 800 Gev $\bar{p}$ has recently been achieved. At DESY 23 GeV $e^+$ on 23 GeV $e^-$ has been reached. LEP will provide $e^+e^-$ at $E_{cm} \sim 90$ GeV at the end of the decade, SLC a little earlier. At DESY, $e^{\pm}$(30 GeV) on $p$(800 GeV) should be achieved in ~1990. We desperately need new experimental results to give direction to our investigations, for we understand physics down to $\sim 10^{-16}$ cm almost too well. It is this understanding that is presented in this book: the resolution of what now seem to be the real mysteries is more deeply buried.

## 1.5 Particles and fields

The subject began in the 1930s, theoretically with the study of relativistic wave equations and experimentally with the discovery of the positron and the muon. The latter particle (a lepton not a meson), was first mistaken for the mediator of the strong nuclear forces postulated by Yukawa. It is actually the second generation electron and we still do not know the answer to Rabi's question "Who ordered that?"

Theoretically, the important first steps were the realisation that antiparticles must exist and the prediction of the existence (and the mass) of the pion. Both came from relativistic wave equations.

For free particles in empty space the wave functions $\psi$ satisfy the Lorentz covariant equation

$$(\nabla^2 - \frac{1}{c^2}\frac{\partial^2}{\partial t^2})\psi = \frac{m^2c^2}{\hbar^2}\psi \tag{1.1}$$

This is the Klein-Gordon equation and is no more than a transcription of the relativistic energy-momentum relation

$$p^2 - \frac{E^2}{c^2} = -m^2c^2, \text{ using } \hat{\mathbf{p}} = \frac{\hbar}{i}\nabla; \ \hat{E} = -\frac{\hbar}{i}\frac{\partial}{\partial t} \tag{1.2}$$

This equation does not in general fully determine $\psi$, but is satisfied by free particles in source-free regions of space. The Dirac equation is first order in $\hat{\mathbf{p}}, \hat{E}$ and describes particles of spin $\frac{1}{2}$. In this case $\psi$ is sufficiently complicated to contain spin as an internal degree of freedom, and can be represented by a 4-component column matrix, but the solutions still satisfy the Klein-Gordon equation.

For any value of $\mathbf{p}$, there are two energy eigenvalues:

$$E = \pm\sqrt{p^2c^2 + m^2c^4} \tag{1.3}$$

The negative energy solutions cannot be neglected. For electrons, the Pauli exclusion principle allows the hypothesis that the negative energy levels of the universe are full, with holes created by promotion to positive energy appearing as positrons. The single particle interpretation (of the Dirac equation) is then lost, and this explanation does not work for bosons. We are forced to reinterpret the negative energy solutions. The ABSORBTION of a negative energy particle (carrying spin, charge, colour...) at a space-time point is interpreted as the CREATION of a positive energy (anti) particle (with opposite spin, charge, colour...) The quantities $\psi$ initially regarded as single particle wave functions must be interpreted as field operators which have the property of creating and annihilating particle states. Here we have the beginning of field theory—50 years old and still going strong.

The solutions of the Klein-Gordon equation are not restricted to free particles. A comparison with electromagnetism is informative. The 4-potentials $A_\mu$ satisfy a similar equation

$$\Box A_\mu = 0 \text{ in empty space } \left[\Box = \nabla^2 - \frac{1}{c^2}\frac{\partial^2}{\partial t^2}\right] \tag{1.4}$$

Picking the scalar potential $\phi$

$$\Box\phi = 0 \tag{1.5}$$

The equation

$$(\Box - \eta^2)\phi = 0 \tag{1.6}$$

is the only second order sourceless generalisation—since $\Box$ is a scalar operator, $\eta^2$ must be a scalar. Thus we have solutions

| | $\Box\phi = 0$ | | $(\Box - \eta^2)\psi = 0$ | |
|---|---|---|---|---|
| $e^{i(k.x-\omega t)}$ | $k^2 - \frac{\omega^2}{c^2} = 0$ | $e^{i(k.x-\omega t)}$ | $k^2 - \frac{\omega^2}{c^2} + \eta^2 = 0$ | Plane waves |
| $\frac{e}{r}^{i(kr-\omega t)}$ | $k^2 - \frac{\omega^2}{c^2} = 0$ | $\frac{e}{r}^{i(kr-\omega t)}$ | $k^2 - \frac{\omega^2}{c^2} + \eta^2 = 0$ | Spherical waves |
| $\frac{q}{r}$ | | $g\frac{e}{r}^{-\eta r}$ | $(\omega \to 0)$ | Static Solution |

where the static solutions imply the existence of a point source. The static solution $\phi$ for a point charge is the electrostatic potential. The static solution $\psi$ may be interpreted as a static potential of range $\frac{1}{\eta}$. The identification of $\eta$ with a quantity $\frac{mc}{\hbar}$ gives the range of the static potential as $\frac{\hbar}{mc}$, the Compton wavelength of the quantum of the free field.

This is how Yukawa got at the pion. Even 50 years ago it was known that nuclear forces are short range, $\sim 10^{-13}$ cm. [If they were not short range the mean nuclear binding energy would grow approximately linearly with the number of nucleons in the nucleus, instead of being approximately constant, ~8 MeV/nucleon]. If these forces are due to fields satisfying a covariant equation, then $\frac{\hbar}{mc}$ ~1fm and hence the free field quantum should have a rest mass ~200 MeV. [We often set $\hbar = c = 1$, when both distance and times are measured in units of $\text{GeV}^{-1}$. Then 1fm= 5.07 ($\text{GeV}^{-1}$). It is a useful exercise to derive this relation]. The pion, the nearest thing to Yukawa's particle, has a mass $\sim$ 140 MeV. This was the original model for the strong interactions, and the pion field is certainly important in the longest range component of the nuclear force.

## 1.6 The strong interactions

The complicated interactions among the hadrons are known collectively as the strong interactions and do not admit a simple description. They are broadly defined by the following characteristics. At high energies, hadron-hadron cross sections are $\gtrsim$ 10 mb [1 barn $= 10^{-24}$ cm$^2$]$\sim (\frac{\hbar}{m_\pi c})^2$. The typical time scale must be $\sim \frac{\hbar}{m_\pi c^2} \sim 10^{-23}$ s (*i.e.* 1fm/c). A large coupling constant is necessary for approximately geometric efficiency. We also note that nuclear binding energies are ~8 MeV/nucleon and even the deuteron is bound by 2.227 MeV. Suppose a Yukawa field were responsible and consider a simple minded way of calculating the binding energy of the hydrogen atom:

$$E = \frac{p^2}{2m_e} - \frac{e^2}{r} \tag{1.7}$$

Use the uncertainty principle to guess a relation $pr = \hbar$ (actually the Bohr condition) when

$$E = \frac{\hbar^2}{2m_e r^2} - \frac{e^2}{r} \tag{1.8}$$

Minimise with respect to $r$:

$$r_{\text{min}} = \frac{\hbar^2}{m_e e^2} \tag{1.9}$$

and

$$E_{\text{min}} = -\frac{1}{2}\frac{e^2}{r_{\text{min}}} = -\frac{1}{2}\left(\frac{e^2}{\hbar c}\right)^2 m_e c^2 \tag{1.10}$$

which is correct. Suppose we neglect the limited range of the Yukawa potential, so that we get the deuteron binding simply by the substitutions $m_e \to M/2$ (reduced mass for two nucleons) and $\frac{e^2}{\hbar c} \to \frac{g^2}{\hbar c}$

Then 2.227 MeV $= \frac{1}{4}\left(\frac{g^2}{\hbar c}\right)^2 Mc^2$

$$\left(\frac{g^2}{\hbar c}\right)^2 \sim 10^{-2} \quad \frac{g^2}{\hbar c} \sim 10^{-1} \quad : \quad \text{compare } \frac{e^2}{\hbar c} = \frac{1}{137}$$

The dimensionless coupling constant must be at least a factor 10 bigger than that for electromagnetism.

It is in fact worse than that. Neglecting the range of the potential, we obtain

$$r_{\text{min}} = \frac{2\hbar^2}{g^2 M} = \frac{2\hbar\ \hbar c}{Mc\, g^2} \sim 20\frac{m_\pi}{M}\left(\frac{\hbar}{m_\pi c}\right) \sim 3\left(\frac{\hbar}{m_\pi c}\right)$$

The exponential which we set equal to unity is in fact $\sim e^{-3}$ at this distance and cannot be neglected. Set

$$E_D = \frac{\hbar^2}{Mr^2} - \frac{g^2}{r} e^{-\frac{m_\pi c}{\hbar} r} \tag{1.11}$$

Set $x = r\frac{m_\pi c}{\hbar}$ and write (1.11) in the form

$$E_D = Mc^2 \left\{ \left(\frac{m_\pi}{M}\right)^2 \frac{1}{x^2} - \frac{g^2}{\hbar c} \left(\frac{m_\pi}{M}\right) \frac{1}{x} e^{-x} \right\}$$

For a bound state to exist at all, $E_D$ must be negative,

so

$$\frac{g^2}{\hbar c} e^{-x} > \frac{m_\pi}{M} \frac{1}{x}$$

for some $x$ or

$$\frac{g^2}{\hbar c} > \frac{m_\pi}{M} \frac{e^{+x}}{x}$$

and $\frac{e^x}{x}$ is a minimum for $x = 1$. Thus for a bound state to exist at all we require $g^2/\hbar c > 0.4$.

A short range potential must have a large coupling in order to overcome the kinetic energy term and produce a bound state.

Nucleon-nucleon forces are not this simple. There are other (even shorter range) meson fields and at distances $\lesssim 1$ fm it now seems likely that nucleon-nucleon forces will be best understood in terms of interacting clusters of coloured quarks. But we now know something about what strong means—a dimensionless coupling constant dangerously close to unity. Is such a coupling constant large enough to produce $\sim$ geometric cross sections? The answer is yes.

## Problems

1.1 Limits on the mass of the electron neutrino $\nu_e$ have been obtained in the laboratory from studies of the spectrum of electrons emitted in tritium $\beta$ decay. From these experiments it is safe to conclude that $m_{\nu_e} < 30$ eV.

In February 1987 the supernova SN1987A appeared in the Large Magellenic Cloud. A few hours before the first sighting, a burst of neutrino interactions was observed in the Kamioka and IMB detectors, instrumented tanks of water set up to search for proton decay. Most, if not all, of the detected interactions were due to $\bar{\nu}_e$. The neutrino energies covered the range 10–40 MeV and the interactions were spread over $\sim$ 10 seconds. Use these observations to estimate an upper limit on the mass of $\bar{\nu}_e$. [Assume that the supernova occurred at a distance of $1.5 \times 10^5$ light years.]

1.2 Consider the effect on the electromagnetic potentials of a finite photon mass $m_\gamma \sim 10^{-15}$eV and calculate the range of such potentials. Try and devise methods of detecting such a finite range. [A detailed discussion may be found in J.D. Jackson *Classical Electrodynamics* 2nd Edition, Wiley 1975.] Calculate the range of the weak interactions responsible for $\beta$ decay, given that $m_{W^\pm} \sim 82$ GeV.

1.3 The root mean square charge radius of the proton is $\sim 0.8$ fm. Estimate the minimum electron laboratory energy necessary to resolve structure on such a scale, assuming a hydrogen target.

The charged pion has a rms charge radius $\sim 0.65$ fm. Calculate the minimum laboratory pion energy necessary to resolve structure on this scale by the scattering of pions from electrons present in a laboratory target.

At HERA, currently under construction beneath Hamburg, 30 GeV electrons will be collided with 800 GeV protons. Estimate the resolution of this most powerful of electron microscopes.

1.4 In an $e^+e^-$ collider such as PETRA, beams are stored for several hours. Throughout that time, both electrons and positrons must pass each turn within $\sim 100\mu$m of the focal points at which detectors are located. Here are some trivial but informative calculations.

(i) Calculate the distance travelled by an electron which survives three hours.

(ii) Calculate the accuracy required—$\sim 100\mu$m divided by the distance travelled in three hours.

(iii) Compare this with the accuracy required to direct a spacecraft to within 20 km at the orbit of Uranus ($3 \times 10^9$ km from the sun).

(iv) In a circular machine, operating at an energy much greater than the rest mass of the particle, the energy loss due to synchrotron radiation is

$$\delta E = \frac{4\pi}{3} e^2 \frac{\gamma^4}{\rho} \qquad \text{(gaussian units)}$$

where $\delta E$ is the energy lost each revolution, $\gamma$ is the ratio of machine energy to particle rest mass energy and $\rho$ is the radius of the machine. Calculate the energy lost each revolution by an electron orbiting in PETRA at 15 GeV, assuming the radius of curvature to be 192m. The circumference of the machine is 2.4km: estimate the rf power which must be supplied to a single electron (in GeV s$^{-1}$) to maintain its energy at 15 GeV. Estimate a lower limit on the power (in MW) consumed when the positive and negative beam currents are each 10ma.

1.5 The luminosity $\mathcal{L}$ of a colliding beam machine is defined by

$$N = \mathcal{L}\sigma$$

where $N$ interactions occur each second for cross section $\sigma$. If such a machine is operating with $n_1$, $n_2$ particles in each bunch in the two beams, which may be assumed to collide head-on, show that the luminosity delivered to one interaction region is

$$\mathcal{L} = \frac{n_1 n_2}{A} f n_B$$

where the cross sectional area of each bunch is $A$, there are $n_B$ bunches in each beam and each bunch makes $f$ complete rotations each second.

The above expression applies for uniform cylindrical bunches. If the bunches have gaussian profiles normal to the beams, with rms coordinates $\sigma_x$, $\sigma_y$, show that the luminosity is given by

$$\mathcal{L} = \frac{n_1 n_2 f n_B}{4\pi\sigma_x\sigma_y}$$

Calculate the luminosity for an $e^+e^-$ collider of circumference 2.4km operated with two bunches in each beam and beam currents of 10ma, if $\sigma_x = 500\mu\text{m}$, $\sigma_y = 50\mu\text{m}$. How many events/hr are accumulated for a cross section of 1nb? [1nb= $10^{-9}$ barns; 1 barn = $10^{-24}\text{cm}^2$.] Calculate the luminosity for a proton synchrotron with a cycle time of 5s delivering $10^9$ particles in each burst to a fixed liquid hydrogen target of length 50cm.

## 2. OF PROPAGATORS AND PIONS.

### 2.1 Potential scattering.

Suppose a strong potential of the form

$$V = (-)\frac{g^2}{r}e^{-\frac{mc}{\hbar}r} \tag{2.1}$$

acts between a pair of hadrons—say nucleons. Let us calculate the scattering cross section due to this potential (we have no machinery at present for calculating more complicated things). In the centre of mass of two nucleons they have equal energy, equal and opposite momenta $p$ and the initial state is taken as a plane wave in the relative separation $\mathbf{r}$:

$$\psi_{in} = e^{i\mathbf{p}.\mathbf{r}/\hbar} \tag{2.2}$$

After scattering, we observe a (fairly) well defined momentum $\mathbf{p}'$: a plane wave in which the direction of the relative momentum has been swung round:

$$\psi_{\text{out}} = e^{i\mathbf{p}'.\mathbf{r}/\hbar} \tag{2.3}$$

$$|\mathbf{p}'| = |\mathbf{p}|$$

$$|\mathbf{p}' - \mathbf{p}| = 2p\sin\frac{\theta}{2}$$

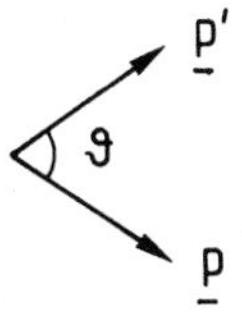


Fig. 2.1

The momentum transfer is $\mathbf{q} = \mathbf{p}' - \mathbf{p}$.

Let the potential $V$ act on the initial state. It generates (acting once) a new state

$$\psi' = V\psi_{\text{in}} \tag{2.4}$$

A specified $\psi_{\text{out}}$ is part of this and we can find out how much we get by expanding $\psi'$ in terms of momentum eigenstates:

$$\psi' = \Sigma a(\mathbf{p}')e^{i\mathbf{p}'.\mathbf{r}/\hbar} \tag{2.5}$$

Then

$$a(\mathbf{p}') \simeq \int e^{-i\mathbf{p}'.\mathbf{r}/\hbar}V(r)e^{i\mathbf{p}.\mathbf{r}/\hbar}d^3r \tag{2.6}$$

and the right hand side of (2.6) is the transition matrix element from the initial state to the final state. It is the three dimensional Fourier transform of the potential with respect to the momentum transfer $\mathbf{q}$

$$\int \psi^*_{\text{out}}V\psi_{\text{in}}d^3r =< \mathbf{p}'|V|\mathbf{p} >= \int e^{-i\mathbf{q}.\mathbf{r}/\hbar}V(r)d^3r \tag{2.7}$$

There is no energy transfer in the centre of mass—that is why we have neglected the time dependence of the nucleon wave functions $\psi$. We can easily evaluate the matrix element

$$\int e^{-i\mathbf{q}.\mathbf{r}/\hbar} V(r) d^3r = \int e^{-iqr\cos\chi/\hbar} V(r) 2\pi r^2 dr d\cos\chi$$

$$= \int \frac{e^{iqr/\hbar} - e^{-iqr/\hbar}}{iqr/\hbar} V(r) 2\pi r^2 dr \tag{2.8}$$

Insert the Yukawa potential (2.1) for $V(r)$:

$$\begin{aligned} < \mathbf{p}'|V|\mathbf{p} > &= \int_0^\infty \frac{e^{iqr/\hbar} - e^{-iqr/\hbar}}{iqr/\hbar} \left(\frac{-g^2}{r}\right) e^{-\frac{mc}{\hbar}r} 2\pi r^2 dr \\ &= \frac{2\pi g^2}{(iq/\hbar)} \left[\frac{1}{\frac{mc}{\hbar} - \frac{iq}{\hbar}} - \frac{1}{\frac{mc}{\hbar} + \frac{iq}{\hbar}}\right] \\ &= \frac{4\pi g^2}{(\frac{q}{\hbar})^2 + (\frac{mc}{\hbar})^2} \end{aligned} \tag{2.9}$$

where $\frac{1}{q^2+m^2}$ is the propagator associated with the field characterised by mass $m$. The transition rate from $\mathbf{p}$ direct to $\mathbf{p}'$ is given by the Fermi Golden Rule. The recipe is obtained from the first Born approximation in scattering theory, or from first order time dependent perturbation theory, as explained in Chapter 7. Take the square of the matrix element, multiply by $2\pi/\hbar$ and the density of final states available. Obtain the cross section by dividing the rate by a flux factor. This is just the relative velocity $2v$ of the two nucleons when we have normalised the wave functions to one per unit volume, as we have chosen to do. Then

$$d\sigma = \frac{1}{2v}\frac{2\pi}{\hbar} | < \mathbf{p}'|V|\mathbf{p} > |^2 \frac{p^2 dp 2\pi d\cos\theta}{dE_{\mathrm{TOT}}(2\pi\hbar)^3} \tag{2.10}$$

$E_{TOT} = 2E$ and $dp/dE = \frac{1}{v}$ (relativistically and non-relativistically). We also have $q^2 = 4p^2 \sin^2 \frac{\theta}{2}$ and since

$$\begin{aligned} d\cos\theta = \sin\theta d\theta &= 4\sin\frac{\theta}{2}\cos\frac{\theta}{2}d\frac{\theta}{2} \\ p^2 d\cos\theta &= 4p^2 \sin\frac{\theta}{2}\cos\frac{\theta}{2}d\frac{\theta}{2} = \frac{1}{2}dq^2 \\ d\sigma &= \frac{1}{4v^2}\frac{2\pi}{\hbar}\left(\frac{4\pi g^2\hbar^2}{q^2+m^2c^2}\right)^2 \frac{\pi dq^2}{(2\pi\hbar)^3} \\ &= \frac{1}{v^2}\pi(g^2)^2 \frac{dq^2}{(q^2+m^2c^2)^2} \end{aligned} \tag{2.11}$$

Then as $v \to c$

$$\sigma = \pi \left(\frac{g^2}{\hbar c}\right)^2 \left(\frac{\hbar}{mc}\right)^2 \tag{2.12}$$

On dimensional grounds alone it could not be anything very different, for the range $\hbar/mc$ is the only scale and $g^2/\hbar c$ is a dimensionless coupling constant. Given that $\hbar/mc \sim 1$ fm, if $g^2/\hbar c \sim 1$ a high energy potential scattering cross section $\sim$ 10mb indeed results. Thus the coupling necessary to produce a bound state also generates cross sections of the right order of magnitude. Beware, however, for we have employed first order perturbation theory and $g^2/\hbar c$ is getting dangerously close to unity...

## 2.2 Virtual particles: annihilation and creation operators

In the calculation of section 2.1, the meson field transferred momentum $\mathbf{q}$ from one nucleon to another. A free particle solution $e^{i(\mathbf{p}.\mathbf{r}-Et)/\hbar}$ transfers momentum $\mathbf{p}$ and energy $E$ from one place to another, subject to the constraint $p^2c^2 - E^2 = -m^2c^4$. In the scattering problem, only a momentum $\mathbf{q}$ has been transferred, subject to no constraint (other than a maximum value from conservation of energy.)

If we observe a real particle, say a pion, it is localised to some extent and we can study the trajectory. Such a real particle cannot be represented by a plane wave with definite momentum—a plane wave is not localised at all. If we accelerate a nucleon (in a high energy collision) and shake off a pion, we actually expand the pion field in its Fourier components when representing it as a plane wave. If the momentum is fairly well defined, the wave packet is narrow in momentum space, and we would represent the amplitude for one component $\mathbf{p}$ by the diagram

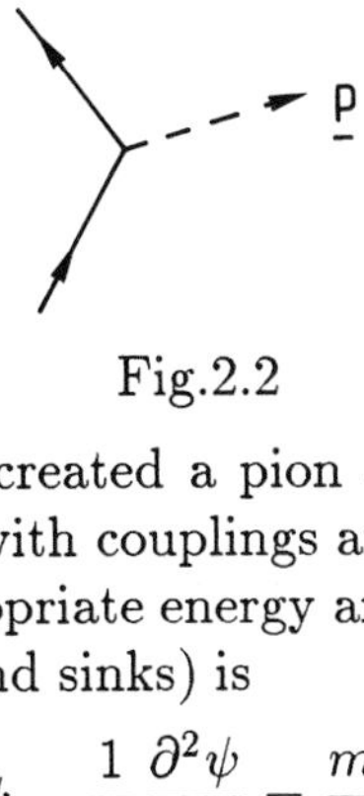


Fig.2.2

where the pion field operator created a pion at the vertex. The Yukawa potential consists of a pion field with couplings and the field operators create and annihilate pions with the appropriate energy and momentum. The general field equation (away from sources and sinks) is

$$\nabla^2\psi - \frac{1}{c^2}\frac{\partial^2\psi}{\partial t^2} = \frac{m^2c^2}{\hbar^2}\psi \tag{2.13}$$

and in the potential problem the time derivative is zero. The matrix element was given by that Fourier component of $\psi$ carrying momentum $\mathbf{q}$ and no energy. The field operators created such a Fourier component at one nucleon and annihilated it at the other. We can represent the scattering process as

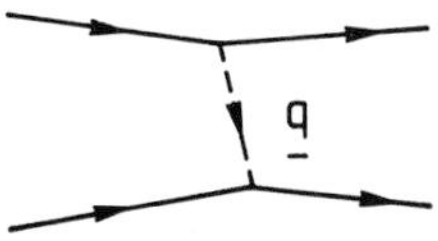


Fig.2.3

The Fourier components not tied to the source are real particles and the Fourier components tied to the source are called virtual particles.

One leg of the scattering process is

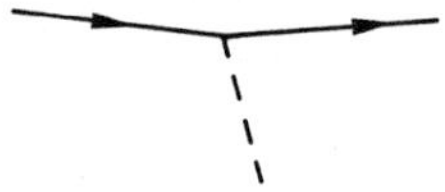

Fig.2.4

At the vertex shown in fig.2.4 we ANNIHILATE a positive energy nucleon with momentum $\mathbf{p}$ and CREATE a positive energy nucleon with momentum $\mathbf{p}'$. In the centre of mass frame the energies are equal and so we CREATE a pion with zero energy and momentum $\mathbf{q} = \mathbf{p} - \mathbf{p}'$. If we do not work in the centre of mass, energy is transferred as well but the matrix element is still $\sim \frac{1}{q^2+m^2}$ where $q^2$ is the square of the 4-momentum transfer (and is the square of the 3-momentum transfer in the centre of mass.)

Suppose however these happy field operators ANNIHILATE a positive energy nucleon ($\mathbf{p}$) and CREATE a negative energy nucleon ($\mathbf{p}'$). This is to be interpreted as ANNIHILATING a nucleon ($\mathbf{p}$) and ANNIHILATING an ANTI-nucleon ($-\mathbf{p}'$)... the created pion field now carries (in the centre of mass) energy $E_{cm} \geq 2M$ and no momentum. The virtual pion would manifest itself as resonance (below threshold) in the $N\bar{N}$ channel:

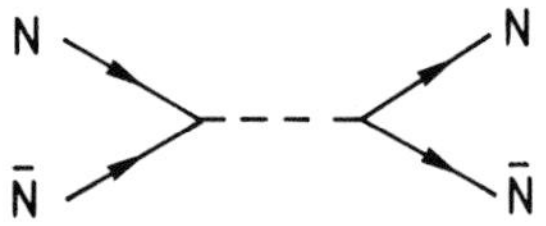


Fig.2.5

It is plausible (and true) that the matrix element is still $\sim \frac{1}{q^2+m^2}$, but in this case $q^2 = -(E_{CM}^2 \geq 4M^2)$. The analogue in non-relativistic quantum mechanics is scattering through an intermediate state.

## 2.3 Charge exchange and isospin

There is a further aspect to the idea of virtual particles. The (free field) pions carry electric charge ($\pi^+, \pi^\circ, \pi^-$) so the annihilation and creation operators also create and annihilate electric charge. The operators retain this property when the field is not free. There is thus a diagram

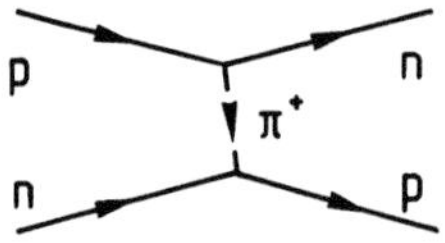


in addition to

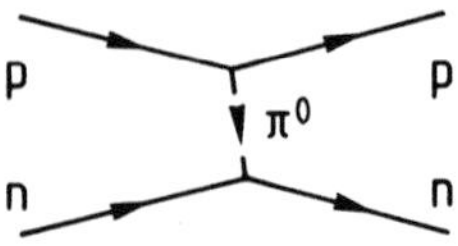


Fig.2.6 Fig.2.7

This charge exchange scattering diagram manifests itself as $np$ backward scattering with cross section $\sim$ $np$ forward scattering. Obviously we can have a neutron turned around and sent back by $\pi^\circ$ exchange, but the virtual $\pi^\circ$ has to carry $q^2 \sim (2p_{cm})^2$ and the matrix element will be smaller than for ($\pi^\circ$) forward scattering. In charge exchange the charge goes across but little momentum: the proton emits a virtual $\pi^+$ which is picked up by the neutron. Something carries charge across, turning neutron into proton and vice versa and the process is easily visualised in terms of virtual charged pion exchange. Charge is discrete and this somehow makes the idea of virtual particle exchange more compelling.

There are very special relationships among the various pion-nucleon couplings. One prominent feature of nuclear forces is charge independence... if electromagnetism is diminished to vanishing point. Take a collection of nucleons in some specified spin-space configuration. Where the Pauli exclusion principle permits, choose protons or neutrons any way you like. (All $pp$ configurations are permitted for $pn, nn$; some $pn$ configurations are forbidden for $pp, nn$). Then the energy of such a collection is independent of the composition.

All two nucleon states can be represented as a product of $\psi_{\text{space}}$ $\psi_{\text{spin}}$ $\psi_{\text{isospin}}$, totally antisymmetric under interchange. The function $\psi_{\text{isospin}}$ takes the form

$$\begin{array}{ll} pp,\ \frac{1}{\sqrt{2}}(np+pn),\ nn & \text{Symmetric} \\ \frac{1}{\sqrt{2}}(np-pn) & \text{Anti-symmetric} \end{array} \tag{2.14}$$

A state accessible to $pp, nn$ is also accessible to $np$ and we CHOOSE the $np$ (isospin) part of this wave function to be the symmetric combination $\frac{1}{\sqrt{2}}(np+pn)$, to go with $pp, nn$. If a meson field is going to impart the same energy to the same spin-space configurations of $pp, np, nn$ and we have charge exchange through the field, then the couplings must be related as follows:

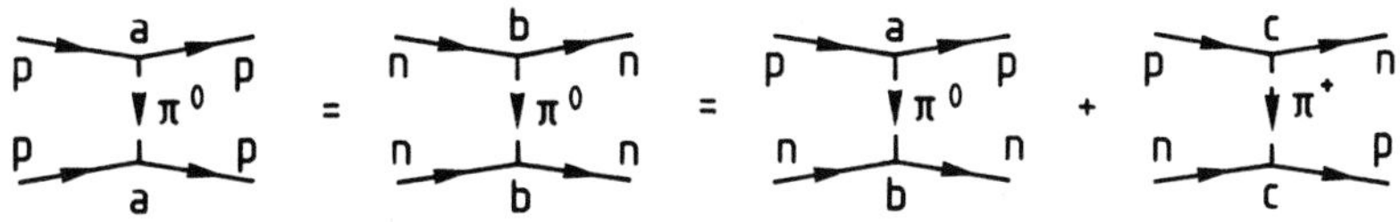


Fig.2.8

where we add these amplitudes because the $np$ states accessible to $pp$, $nn$ are symmetric under $n \leftrightarrow p$ —which is a physical operation. There is a trivial solution: $a = b$; $c = 0$. The interesting one is $a = -b$; $c = \pm\sqrt{2}b$. We choose

$$\begin{array}{lll} a = & -g\sqrt{\frac{1}{3}} & p \to p\pi^\circ \\ b = & g\sqrt{\frac{1}{3}} & n \to n\pi^\circ \\ c = & g\sqrt{\frac{2}{3}} & p \to n\pi^+ \end{array} \tag{2.15}$$

Two phases are arbitrary and the normalisation is such that

$$(p \to p\pi^\circ)^2 + (p \to n\pi^+)^2 = g^2 \tag{2.16}$$

These numbers are just the appropriate Clebsch-Gordan coefficients for SU(2) or angular momentum $\frac{1}{2} \to 1 \otimes \frac{1}{2}$. The pion is an isospin triplet (the 1 in $1\otimes\frac{1}{2}$). The arbitrary phases have been chosen so as to conform with the Condon and Shortley convention which is standard for tables of angular momentum coefficients.

The existence of charge exchange potentials (or amplitudes) gives physical reality to the isospin functions $\frac{1}{\sqrt{2}}(np \pm pn)$ which are symmetric and antisymmetric under swapping charges between proton and neutron. This notation is more than simple book keeping, because the characteristics really do get interchanged.

## 2.4 Resonances

We can play these games in other ways. Nucleon-pion scattering will have a contribution from the diagrams through a virtual nucleon

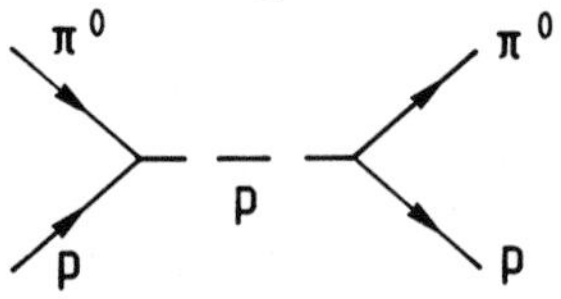


Fig.2.9

$\pi^+ p$ scattering will not take place this way of course, but there is a contribution from isospin $\frac{3}{2}$ amplitudes, for example

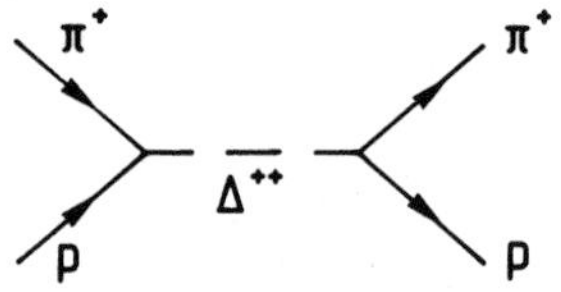


Fig.2.10

which appears as an elastic $J = \frac{3}{2}$ resonance in $\pi^+ p$ scattering—scattering through an intermediate state or particle—where the resonance peak is at about 1.23 GeV and the full width at half maximum is $\sim$ 100 MeV. Using $\Gamma\tau = \hbar$, $\tau \sim 10^{-23}$s as expected for these strong interactions. To keep charge independence between nucleons, this state must be duplicated in the other $\pi N$ channels... there are 4 possible charge states, $\Delta^{++}, \Delta^{+}, \Delta^{\circ}, \Delta^{-}$, an isospin quartet. Clearly $p\pi^+, n\pi^-$ are pure $T = \frac{3}{2}$ states. But $p\pi^\circ, n\pi^+$ can be $T = \frac{1}{2}$ or $\frac{3}{2}$. Since

$$p \to -\sqrt{\frac{1}{3}}p\pi^\circ + \sqrt{\frac{2}{3}}n\pi^+$$

and $p$ is $T = (\frac{1}{2}, +\frac{1}{2})$

$$T = (\tfrac{1}{2}, +\tfrac{1}{2}) = -\sqrt{\frac{1}{3}}p\pi^\circ + \sqrt{\frac{2}{3}}n\pi^+$$

and then

$$T = (\tfrac{3}{2}, +\tfrac{1}{2}) = \sqrt{\frac{2}{3}}p\pi^\circ + \sqrt{\frac{1}{3}}n\pi^+$$

the orthogonal combination. Similarly $n$ is

$$T(\tfrac{1}{2}, -\tfrac{1}{2}) = \sqrt{\frac{1}{3}}n\pi^\circ - \sqrt{\frac{2}{3}}p\pi^-$$

and so

$$T = (\tfrac{3}{2}, -\tfrac{1}{2}) = \sqrt{\frac{2}{3}}n\pi^\circ + \sqrt{\frac{1}{3}}p\pi^-$$

Thus

$$\begin{aligned}
(\tfrac{3}{2}, +\tfrac{3}{2}) &= (\tfrac{1}{2}, +\tfrac{1}{2})(1, +1) \\
(\tfrac{3}{2}, +\tfrac{1}{2}) &= \sqrt{\frac{2}{3}}(\tfrac{1}{2}, +\tfrac{1}{2})(1, 0) + \sqrt{\frac{1}{3}}(\tfrac{1}{2}, -\tfrac{1}{2})(1, +1) \\
(\tfrac{3}{2}, -\tfrac{1}{2}) &= \sqrt{\frac{1}{3}}(\tfrac{1}{2}, +\tfrac{1}{2})(1, -1) + \sqrt{\frac{2}{3}}(\tfrac{1}{2}, -\tfrac{1}{2})(1, +1) \\
(\tfrac{3}{2}, -\tfrac{3}{2}) &= (\tfrac{1}{2}, -\tfrac{1}{2})(1, -1)
\end{aligned} \tag{2.17}$$

keeping the Condon and Shortley convention for arbitrary phases again. These can be inverted to give the isospin composition of $p\pi^\circ, n\pi^\circ, p\pi^-, n\pi^+$ of course. The physical significance of these mathematical relationships is elucidated in Problem 2.2.

The $\Delta^{++}$ was the first of the resonant states in the strong interactions to appear—only the first, there are probably an infinite number of them. The cosy little world of nucleons and pions began to take on all the reality of fairyland. Even the pion field is not so simple as our treatment has implied so far... the pion is a pseudoscalar, not a scalar. It couples to the nucleon through a term $\boldsymbol{\sigma}.\boldsymbol{\nabla}\phi$ (nonrelativistically; relativistically this is a Dirac $\gamma_5$ coupling) and this gives rise not only to a central force but also to the non-central spin-spin, spin-orbit and tensor forces. All are present in the long range component of the nucleon-nucleon force, which is an indication that the pion is pseudoscalar. The parity of the pion has been directly determined from the observation of the process $\pi^- d \to nn$.

The pion has spin zero and is captured from an $s$-state on the spin 1 deuteron. Therefore the final state $nn$ has total angular momentum 1. The Pauli principle prevents $nn$ being in even orbital angular momentum states with $J = 1$, so it must be a $p$-wave, which is odd under coordinate reflection. Thus the final state has negative parity, for the deuteron is (mostly) $s$-wave. Hence for parity conservation (true for the strong interactions) the pion has an intrinsic negative parity. This was originally envisaged as a possible, but mysterious property of the pion field. We account for it now trivially. The pion is an $s$-wave singlet state of $q\bar{q}$—and with a fermion-antifermion pair is associated a negative parity. We need the Dirac equation (Ch.8) to obtain this result. Incidentally, we know the spin of the $\pi$ various ways. First, the decay $\pi^\circ \to \gamma\gamma$ means $J_\pi = 0, 2, 3, 4\ldots$ (not obvious). However you make it, $\gamma\gamma$ is isotropic and therefore the spin of the pion is spin 0.

Secondly, the relative cross sections for the reactions $p + p \leftrightarrow \pi^+ + d$ at the same centre of mass energy are dependent on the pion spin. This argument is not simple either... although it is often pretended that it is, with the insertion of some phrase such as "Detailed balance implies..." Great care and a clear head are necessary—see Problem 2.5(i).

## Problems

2.1 The cross sections for neutrino interactions are of the order of $10^{-44}\mathrm{cm}^2$ at a laboratory neutrino energy of $\sim 1$ MeV. Show that if such cross sections are attributed to a potential associated with the $W$ bosons of mass $\sim 80$ GeV, then the dimensionless coupling $g_W^2/\hbar c$ has a value $\sim 10^{-2}$, where $g_W$ is the coupling of the participating fermions to the $W$ boson field.

2.2 We deduced the relationships among the pion-nucleon couplings by comparing the lowest order nucleon-nucleon scattering amplitudes and requiring that these amplitudes satisfy charge independence. Consider the second order diagrams in the class

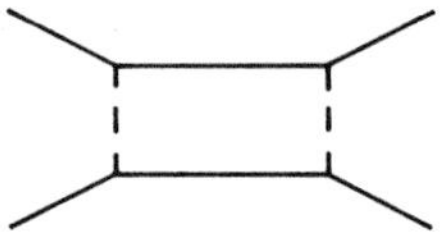

(where the solid lines represent nucleons and the broken lines exchanged pions) and show that these second order amplitudes also satisfy charge independence, using the pion-nucleon couplings deduced from the first order diagrams.

2.3 Another class of second order diagrams can be represented by

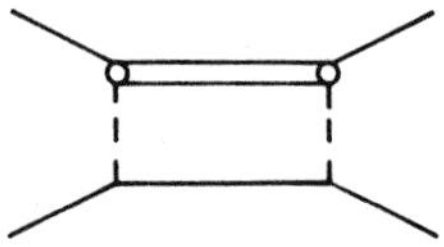

where the solid lines are nucleons and the double line represents one of the four $T = \frac{3}{2}$ $\Delta$ states. Show that the amplitudes summarised by this figure again satisfy nucleon-nucleon charge independence, provided that the $\pi N \Delta$ amplitudes satisfy the relations implied by (2.17).

2.4 Show that, given charge independence of nuclear forces, the bound states of two nucleons in a given spatial configuration form a degenerate isospin triplet and an isospin singlet which in general has different energy. [In fact only the $1^+pn$—the deuteron—is bound.]

Suppose that there were to exist three kinds of nucleon with a generalised charge independence of the forces. What multiplet structures would arise in two nucleon systems? [This can be figured out from basic physical principles: it is neither necessary nor advantageous to resort to group theory. In the mid 1950s it seemed possible that $p$, $n$, $\Lambda$ might be such a basic triplet.]

2.5* Here are two problems concerning the spin of the pion.

(i) The ratio of the cross sections for the processes

$$\pi^+ d \to pp; \qquad pp \to \pi^+ d$$

(at the same centre of mass energies) is given by

$$\frac{\sigma(pp \to \pi^+ d)}{\sigma(\pi^+ d \to pp)} = \frac{(2s_d + 1)(2s_\pi + 1)}{\frac{1}{2}(2s_p + 1)^2} \frac{p_\pi^2}{p_p^2}$$

where $p_\pi$, $p_p$ are the centre of mass momenta of the pion and one of the protons respectively. The crucial ingredient in the derivation of this formula is that the squared matrix elements $| < a|H|b > |^2$ and $| < b|H|a > |^2$ are equal (at the same centre of mass energies), where $b$ and $a$ denote completely specified states. Show that the above ratio of total cross sections is obtained by injecting this equality into the Fermi Golden Rule and performing an average over the spin states of the unpolarised projectiles and targets.

(ii) A final state consisting of two photons can be specified in the centre of mass by the relative momentum vector and either two polarisation vectors giving the direction of the electric field associated with the two photons or by the helicities of the two photons (which is equivalent to specifying the circular polarisation state of each). The restrictions on the angular momentum of such a state come from the requirements of Bose symmetry and transverse (linear) polarisation. A complete treatment was first given by C.N. Yang, *Phys.Rev.* **77** 242 (1950).

It is easy to write down a scalar function which satisfies these conditions and not much harder to write down a pseudoscalar function. Two photons can therefore exist in a state of zero angular momentum and either positive or negative parity. Find those functions. Then search for a suitable function which transforms under rotations as a vector. You will find that two real photons cannot exist in a state of spin 1.

# 3. STRANGERS IN TOWN.

## 3.1 The muon.

The muon was discovered shortly after Yukawa's work, in cloud chamber photographs of the penetrating component of the cosmic radiation. The momenta of such tracks were obtained from curvature in an applied magnetic field; velocities were estimated from the density of droplets along the trail. The mass was thus established as $\sim 200 m_e$ and it was natural that it should be taken as Yukawa's particle. It is not, but its parent the pion is [$m_\mu = 106$ MeV; $m_\pi = 140$ MeV].

First, the muon is weakly interacting. Determine the lifetime of $\mu^+$ by stopping them in matter and measuring the interval before decay. The mean is $2 \times 10^{-6}$s. Do the same with $\mu^-$. Now $\mu^+$ is kept away from nuclei by Coulomb repulsion (as is $\pi^+$) but $\mu^-$ is captured into the lowest ($\mu$) Bohr orbit in $\approx 10^{-12}$s. The timescale for the strong interactions is $\approx 10^{-23}$s and this could not be diluted much by the ratio of the (muonic) atomic volume and the volume of the nucleus. The mean interval between stopping $\mu^-$ and observing its decay is $\sim 2 \times 10^{-6}$s in light elements such as carbon and only for high $Z$ nuclei does capture dominate. Negative muon capture is a weak process not a strong one. The negative pion is captured fast and the positive pion, repelled from nuclei, decays — into a $\mu^+$ and an invisible particle, the muon neutrino. The sequence of pion production through the strong interactions, decay of the positive pion to muon and subsequent decay of the muon to an electron was elucidated in a few years following the first observations in 1947. The observations were made with photographic emulsion in which ionisation caused by charged particles sensitises grains of silver bromide. The tracks in the developed plates yielded information about energy (from the distance a particle travelled), velocity (from the density of developed grains along the track) and momentum (from multiple scattering). Even in the early days of photographic emulsion exposed to the cosmic radiation, it soon became clear that the muon following pion decay at rest is monoenergetic, but the electron from muon decay is not. A decade later the complete sequence became a common observation in the heyday of bubble chamber operation.

Secondly, the muon has spin $^1/_2$, and is a weakly interacting fermion, a lepton, and not a meson at all (in modern terminology). How do we know the spin? The decay rate and electron spectrum are wholly consistent with

$$\mu^+ \to e^+ \bar{\nu}_\mu \nu_e$$

and the standard theory of the 4-fermion weak interaction. The electron spectrum agrees with that expected for a massless $\nu_\mu$; the upper limit is $m_{\nu_\mu} < 0.25$ MeV. You can precess the spin of the muon in a magnetic field (the decay electrons are aligned preferentially relative to the muon spin because the weak interactions do not conserve parity) and the g-factor is 2.002.... The fine structure of muonic atoms and the hyperfine structure of muonium ($\mu^+ e^-$ bound state) have been studied. The muon is just a heavy electron, beautifully described by the Dirac equation (+ QED corrections).

There is another funny feature about the muon, apart from its existence as an (apparently) superfluous copy of the electron:

$$\mu \not\to e\gamma \quad \text{(branching ratio } < 1.7 \times 10^{-10}\text{)}$$
$$\not\to ee^+e^- \quad \text{(branching ratio } < 2.4 \times 10^{-12}\text{)}$$

Furthermore, with beams of neutrinos made in association with muons

$$\pi^+ \to \mu^+ \nu_\mu \quad ; \quad \pi^- \to \mu^- \bar{\nu}_\mu$$
$$\nu_\mu + n \to \mu^- + p \quad \text{but not} \quad e^- + p$$
$$\bar{\nu}_\mu + p \to \mu^+ + n \quad \text{but not} \quad e^+ + n$$

There is a quality of muonness which seems to be conserved.

### 3.2 Strange particles.

Among the hadrons there is something similar but perhaps even funnier. In the early 1950s particles were observed, in the debris from interactions of the cosmic radiation, with lifetimes $\sim 10^{-8} - 10^{-10}$ seconds and yet present in such profusion that they must be made by strong interactions. The decay products are pions and nucleons, strongly interacting particles. It would therefore be expected that the lifetimes should be characteristic of the strong interactions, $\sim 10^{-23}$s. In fact the lifetimes are characteristic of the weak interactions, responsible for nuclear $\beta$ decay. This is not quite obvious, for the neutron has a lifetime $\tau_n \sim 10^3$s, but we can make it so. In the simple Fermi theory $\beta$ decay takes the form of a 4-fermion point interaction

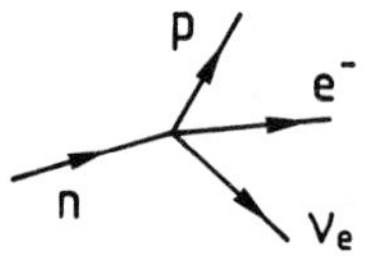


Fig.3.1

and using the Fermi Golden Rule

$$\frac{1}{\tau} = \frac{2\pi}{\hbar}|\mathcal{M}|^2\rho \tag{3.1}$$

where $\rho$ is phase space. For three particles in the final state, under the standard kinematic assumptions of nuclear $\beta$ decay

$$d\rho \sim \frac{p^2 dp (E_0 - E)^2}{(2\pi\hbar)^6 c^3} \quad \text{so} \quad \int d\rho \sim E_0^5 \tag{3.2}$$

where $p$, $E$ are the electron momentum and energy. In nuclear $\beta$ decay, the matrix element involves a coupling constant and a nuclear overlap integral. Neutron and proton are the same animal and so for neutron decay

$$\frac{1}{\tau_n} \sim G^2 E_0^5 \tag{3.3}$$

In this simple theory $G$ has dimensions [erg cm$^3$ or, in natural units $\hbar = c = 1$, GeV$^{-2}$]. If we have a genuine weak decay process for these long lived hadrons ($K^0 \to \pi^+\pi^-$, $\Lambda^0 \to p\pi^- \ldots$ which do not look like 4-fermion interactions) in the sense that the coupling constant is $G$, then we must still have

$$\frac{1}{\tau} \sim G^2 E_0^5 \tag{3.4}$$

and $E_0$ must be approximately the particle mass $M$, for decay into much lighter particles. Then

$$\frac{\tau_M}{\tau_n} \sim \left(\frac{E_{0n}}{M}\right)^5 \quad \begin{array}{l} E_{0n} \sim 1.5 \text{ MeV for neutron decay} \\ M \approx 0.5 \text{ GeV for strange particles} \end{array} \tag{3.5}$$

so $\tau_M \approx 10^{-9}$s.

It does not matter that most of the final states from decay of these strange particles are two body. The extra dimensional factors must come from somewhere in the internal machinery, and if the interaction is a weak interaction and there is no freak cancellation, then $\frac{1}{\tau_M} \sim G^2 M^5$.

This works very well for muon decay — which is a 4-fermion interaction with three free particles in the final state, for setting $M = m_\mu = 106$MeV

$$\tau_\mu/\tau_n \sim 10^{-9} \text{ and } \tau_\mu \sim 10^{-6}\text{s}$$

[In fact nuclear $\beta$ decay through the vector (Fermi) interaction (e.g. $^{14}$O) is about 5% slower than would be inferred from muon decay ($\cos^2\theta_C$) and $\beta$ decay of strange particles is only about 5% of the expected rate ($\sin^2\theta_C$).]

Thus there are strongly interacting particles which decay to strongly interacting particles only weakly. Something stops them decaying strongly. These strange strongly interacting particles carry a new internal quantum number called strangeness which is conserved in strong and electromagnetic interactions, but not in the weak interactions. Strange particles are made in pairs of zero total strangeness.

$$\begin{array}{lll} \pi^- p \to & K^0 & \Lambda^0 \\ & S = +1 & S = -1 \end{array}$$

How might we make them? Perhaps through a resonance

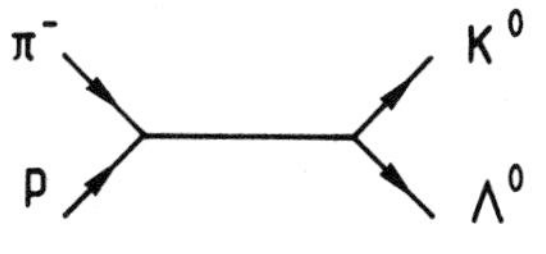


Fig.3.2

or perhaps by strangeness exchange.

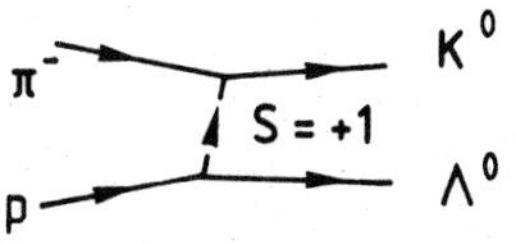


Fig.3.3

Heavier strange particles could decay strongly into lighter strange particles with widths ~100MeV, $\tau \sim 10^{-23}$s. They would show up as resonances in such channels as $K^-p$, $K\pi$... and they do. For example

$$K^*(890) \rightarrow K\pi \quad \Gamma = 50\text{MeV}$$
$$\Sigma(1385) \rightarrow \Lambda\pi \quad \Gamma = 35\text{MeV}$$

(but $\Omega^-$ (1672) has $S = -3$ and is not quite massive enough to decay to $\Xi(S = -2)K$).

It is interesting and curious that while mesons have both positive and negative strangeness, baryons exist only with negative strangeness and antibaryons only with positive strangeness.

The hypothesis of strangeness, conserved except in the weak interactions, does little more than give a curious phenomenon a name. It introduced order into our ignorance and at least the prediction of associated production could be tested, and was rapidly confirmed once accelerators disposing of proton beams of energy > a few GeV came on the air. Proton synchrotrons with beam energies ~ 2–30GeV were all we had until the mid 70s and a huge amount of data on the multitude of hadrons was built up. We now know that strange particles are strange because they contain strange quarks ... but we still do not know what strangeness (or muonness) really is.

### 3.3 Peculiarities of the weak interactions.

Have a further look at the peculiarities of the weak interactions.

(i) Mean lifetimes vary as $\tau_M \sim 10^{-12}M^{-5}$s ($M$ in GeV). For $M$ (only) ~ 100GeV the decay rate is approximately that appropriate to the strong interactions. This result is directly related to

(ii) The coupling constant $G^2$ has dimensions GeV$^{-4}$ in natural units. We cannot construct a dimensionless number from this without inserting a mass scale. In natural units $GM^2$ is dimensionless and $(GM^2)^2$ might be taken as the dimensionless constant analogous to $\alpha$ $(e^2/\hbar c)$. The only problem is, what is $M$?

With the standard definition of the weak coupling constant, careful analysis of nuclear $\beta$ decay and muon decay gives

$$G_F = 1.17 \times 10^{-5}\text{GeV}^{-2} \quad (1.4 \times 10^{-49} \text{ erg cm}^3)$$

$$\begin{aligned} \text{If the scale is the proton mass} \quad & (GM_p^2)^2 \sim 10^{-10} \\ \text{electron mass} \quad & (Gm_e^2)^2 \sim 10^{-23} \\ \text{but if the scale is } M_W \sim 100\text{GeV} \quad & (GM_W^2)^2 \sim 10^{-2} \approx \alpha \end{aligned}$$

A choice of scale such as $M_p^2$ is arbitrary (and is physical nonsense). The correct scale is $M_W \sim 100$GeV.

The low energy theory cannot be correct. The cross section for $\nu_\mu + n \rightarrow \mu^- + p$ (pretending that $p, n$ are pointlike) or better, true elastic scattering of pointlike particles

$$\nu_e + e \rightarrow \nu_e + e$$

takes the form (Problem 3.3)

$$\sigma = \frac{2\pi}{\hbar}\frac{1}{2c}G^2\frac{4\pi}{(2\pi\hbar)^3}p^2\frac{dp}{dE_{cm}} \tag{3.6}$$

in the simple theory. This is constructed by having the weak interaction act once between an initial and a final plane wave. In natural units ($\hbar = c = 1$)

$$\sigma = \frac{G^2}{4\pi}p^2 \quad p = E_{cm}/2 \tag{3.7}$$

With the full weak interaction matrix element and the conventional definition of $G_F$† the result is

$$\sigma = \frac{G_F^2}{\pi}s \quad s = E_{cm}^2 \tag{3.8}$$

For the assumed pointlike interaction, only waves which do not vanish at the origin take part in the interaction and these are only a little bit of a plane wave. Split a simple plane wave up into a superposition of incoming and outgoing spherical waves (partial wave analysis)

$$e^{i\mathbf{k}.\mathbf{r}} = e^{ikr\cos\theta} = \sum A_l(r)P_l(\cos\theta) \tag{3.9}$$

For $l = 0$

$$\int_{-1}^{+1} e^{ikr\cos\theta}P_0(\cos\theta)d\cos\theta = \sum A_l(r)\int_{-1}^{+1} P_l(\cos\theta)P_0(\cos\theta)d\cos\theta \tag{3.10}$$

So

$$\frac{2\sin kr}{kr} = 2A_0(r) \tag{3.11}$$

This amplitude is finite at the origin. For higher $l$ the centrifugal barrier pushes the wave function away from the origin, and $A_l(r) \sim (kr)^l$ for $kr \ll 1$. Only the lowest angular momentum states can contribute to a pointlike interaction. For an elastic scattering process the total wave function is given by

$$\psi_{\text{tot}} = \frac{e^{ikr}e^{2i\delta} - e^{-ikr}}{2ikr}\left[\times e^{-i\omega t}\right] \tag{3.12}$$

because by causality we can only interfere with the outgoing piece.

$$\psi_{sc} = \psi_{\text{tot}} - \psi_{in} = \frac{e^{ikr}(e^{2i\delta} - 1)}{2ikr} \tag{3.13}$$

The scattering cross section is the integrated outgoing flux divided by the incident flux (just like Thomson scattering of soft X-rays from free electrons) and this is

$$\sigma_0 = \frac{4\pi}{k^2}\sin^2\delta \le \frac{4\pi}{k^2} \tag{3.14}$$

† See Chs.8 and 9.

if we just shift the phase in true elastic scattering, or

$$\sigma_0 \leq \frac{\pi}{k^2} \tag{3.15}$$

if we absorb all the outgoing part by converting it into something else. We cannot exceed such limits unless the interaction point injects probability (and energy) not present in the incident wave; this is a unitarity limit. Note that $k$ is the momentum of one particle (in the centre of mass) in natural units and so

$$\sigma_0 = \lesssim \frac{4\pi}{s} \tag{3.16}$$

for a pointlike interaction. Compare the two results (3.8) and (3.16):

| Weak theory | Unitarity limit |
|---|---|
| $\sigma = \dfrac{G_F^2 s}{\pi}$ | $\sigma \lesssim \dfrac{4\pi}{s}$ |

The above unitarity limit was obtained for a plane wave representing spinless particles. The wavefunction for two relativistic spin 1/2 particles is more complicated, but these expressions must be right (up to numerical factors) simply on dimensional grounds. We have a contradiction under the most optimistic assumptions if

$$G_F^2 s^2 \approx (4\pi)^2$$

and $G_F^2 s^2 \approx 1$ when $s = 10^5\text{GeV}$, $E_{cm} = 300\text{GeV}$.

In fact $\sigma(\nu_\mu + n \to \mu^- + p)$ ('elastic') flattens out for laboratory $\nu_\mu$ energies > a few GeV†. This is because hadrons are not pointlike. But the total cross section $\sigma(\nu_\mu + n \to \mu^- + X)$ continues rising linearly with laboratory energy (i.e. with $s$) up to at least 100GeV, and the problem remains for $\nu_e + e \to \nu_e + e$. Either $\nu_e, e$ are not pointlike (we know that electrons are pointlike to below $10^{-16}$cm because processes such as $e^+e^- \to e^+e^-$, $\mu^+\mu^-$ agree with pointlike QED for $q^2 \sim 1000\ \text{GeV}^2$) or the interaction is not pointlike. The obvious way out of all these peculiarities of the Fermi theory is to replace

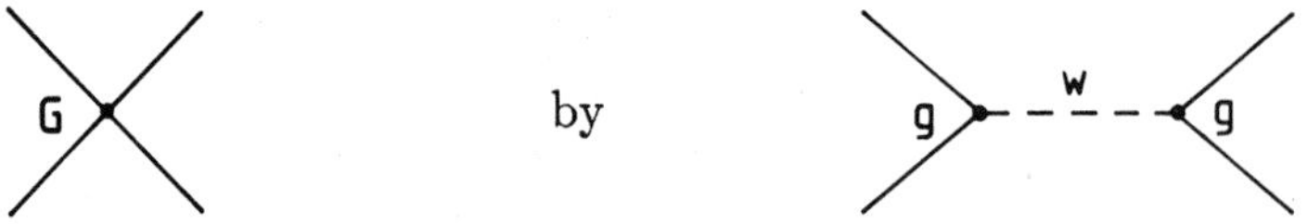


Fig.3.4

where $M_W$ must be $\lesssim 100\text{GeV}$. (This was another idea of Yukawa, but he economically hoped that the pion would do the job). Then the Intermediate Boson propagator $\frac{1}{q^2+M_W^2}$, where $q$ is the 4-momentum carried by the virtual boson, will control the growth of cross sections and we interpret (see Ch.2)

$$G \to \frac{4\pi g^2}{q^2 + M_W^2} \tag{3.17}$$

† See Ch.6.

At low energies, $q^2 << M_W^2$ this yields

$$G \to \frac{4\pi g^2}{M_W^2} \tag{3.18}$$

The natural scale is thus $M_W^2$ and with $M_W \approx 100\text{GeV}$ $g^2 \simeq \alpha$, the electromagnetic coupling (see Problem 2.1). This is rather nice, since both the photon and the $W$ have spin 1. (The $W$ must have spin 1 because otherwise allowed Gamow-Teller interactions are impossible in nuclear $\beta$ decay). Yet another peculiarity remains. The weak interactions violate parity to the maximal possible extent (and electromagnetism does not). We do not understand this, although we have successfully described the phenomenon. When constructing the (very successful) intertwined electro-weak theory, parity violation is written in from the start. The $W$ bosons couple only to left handed fermions and right handed anti-fermions (this is more than just the existence of the left handed neutrino) and so the weak interactions violate not only parity but also charge conjugation (particle-antiparticle) invariance.

## 3.4 The interactions of particle physics.

We conclude this chapter by summarising the relevant properties of the interactions of particle physics, to the extent that we have studied them so far.

The electromagnetic interaction has infinite range, corresponding to the zero rest mass of the photon, and the dimensionless coupling $e^2/\hbar c$ ($e^2$ in natural units) is $\sim \frac{1}{137}$. There are four potentials, the scalar potential $\phi$ and the magnetic vector potential $\mathbf{A}$, which together make a four-vector. The photon has spin 1 (but as a consequence of the transverse conditions $\nabla.\mathbf{E} = 0$, $\nabla.\mathbf{B} = 0$ the only spin states are along or against the direction of motion). Because of the infinite range, the classical limit is familiar.

The term strong interaction covers those interactions among hadrons which are characterised by cross sections $\sim$ 10mb and timescales $\sim 10^{-23}$s. There is no fundamental field: the hadrons are quark composites. Effective field theories are relevant on a scale $\sim 10^{-13}$cm ($m_\pi \simeq 140$ MeV) and this is the range of the strong interactions. The dimensionless coupling is $\approx 1$. (The interactions among the quarks—QCD—are discussed in Chs.5 and 11–13).

The weak interactions we have considered are charge exchange processes. The field quanta have mass $M_W \sim 100$ GeV and the corresponding range of the field is $\sim 2 \times 10^{-3}$fm. The dimensionless coupling between fermions and the intermediate vector boson is $\sim e^2$. The fields couple only to the left handed components of fermions and the interaction violates parity and charge conjugation invariance to the maximum possible extent.

There is another infinite range force with a familiar classical limit: gravitation. The source is the energy-momentum density tensor rather than a (four-vector) current density and the quanta would have spin 2. Gravitation can apparently be completely neglected at the femtoscopic scale.

## Problems

3.1 Use the Fermi Golden Rule to estimate the partial lifetime against capture for a $\mu^-$ in the ground state of i) muonic carbon ($Z = 6$) ii) muonic iron ($Z = 26$). The basic process is

$$\mu^- + [p] \rightarrow \nu_\mu + n$$

where the proton $[p]$ is bound in the nucleus. [It is relevant to calculate the recoil energy of a free neutron.]

3.2 Calculate the kinetic energy of a muon from pion decay at rest. Estimate the range of such a muon in i) a photographic emulsion which is 50% gelatine, 50% silver bromide by volume ii) in a liquid hydrogen bubble chamber.

[The ionisation loss of a charged particle in matter is $\sim 2$ MeV/gm/cm$^2$ for $v \sim c$ except in $H_2$. What is the equivalent number for $H_2$, and how does ionisation loss scale with velocity?]

3.3 It is trivial to obtain eq.(3.6) from the Fermi Golden Rule—so do it. Estimate neutrino-nucleon cross sections for laboratory energies of (i) 1 MeV (ii) 1 GeV (iii) 100 GeV. Calculate the fractions of neutrinos that interact passing through three metres of liquid deuterium.

3.4 Calculate the partial decay rate $W^- \rightarrow e^-\bar{\nu}_e$, assuming the dimensionless coupling $g_W^2 \sim e^2$. Hence estimate the total width of the $W^-$ (in GeV) assuming that it decays into $e^-\bar{\nu}_e$, $\mu^-\bar{\nu}_\mu$, $\tau^-\bar{\nu}_\tau$ and three colours of $(\bar{u}d)$, $(\bar{c}s)$. [$M_{W^-} \sim 80$ GeV.]

3.5 At what centre of mass energy would the gravitational interaction between two electrons (approximately) equal the electromagnetic interaction?

Consider what limits can be placed on the mass of the graviton.

## 4. HADRONS.

### 4.1 Regularities among hadrons

There are two remarkable regularities among the hadrons, which became apparent in the early 1960s. The first is that the light hadrons (those composed of *uds* quarks) fall into groups of multiplicity 1, 8, 10 (compare with isospin multiplets 1, 2, 3, 4.) The mesons come in 1s and 8s, baryons in 1, 8, 10 and there are no positive strangeness baryons. The second regularity is that states which can be interpreted as rotational excitations lie on straight lines of angular momentum versus mass squared and all straight lines have the same slope.

The lightest hadrons are:

| MESONS | | | | | | BARYONS | | | | | |
|---|---|---|---|---|---|---|---|---|---|---|---|
| T | S | | $\leftarrow T_3$ | | M (MeV) | T | S | | $\leftarrow T_3$ | | M (MeV) |
| $\frac{1}{2}$ | +1 | $K^+$ | | $K^\circ$ | 495 | $\frac{1}{2}$ | 0 | $p$ | | $n$ | 938 |
| 1 | 0 | $\pi^+$ | $\pi^\circ$ | $\pi^-$ | 140 | 1 | $-1$ | $\Sigma^+$ | $\Sigma^\circ$ | $\Sigma^-$ | 1190 |
| 0 | | | $\eta, \eta'$ | | | 0 | | | $\Lambda^\circ$ | | |
| $\frac{1}{2}$ | $-1$ | $\bar{K}^\circ$ | | $K^-$ | (495) | $\frac{1}{2}$ | $-2$ | $\Xi^\circ$ | | $\Xi^-$ | 1320 |
| | | | $8+1$ | | | | | | 8 | | |
| | | | $J^P = 0^-$ | | | | | | $J^P = \frac{1}{2}^+$ | | |

The charged pions and the kaons were discovered in studies of the cosmic radiation using photographic emulsion and cloud chambers. The $\pi^\circ$ was discovered from the characteristic energy spectrum of photons emitted following absorbtion of $\pi^-$ at rest on protons. Both $\eta$ and $\eta'$ were found in the bubble chamber era. Among the baryons, the $\Lambda$ was the first strange baryon to be observed, in cloud chamber pictures, and representatives of the $\Sigma$ isospin triplet and the $\Xi$ (cascade) doublet were also first found in the cosmic radiation. Those heroic days came abruptly to an end with the commissioning of the first proton synchrotons which could produce strange particles.

### 4.2 Determination of hadron properties.

We shall not say much about the determination of the basic properties of mass and lifetime. Isospin is determined from the number of members of a multiplet and also by comparing cross sections and branching ratios with isospin coupling coefficients such as those encountered in Ch.2. The determination of spin and parity is relatively complicated.

In general there are methods internal and methods external and such methods apply to the strongly decaying (resonant) states as well. For the ground state baryons above, violation of parity in the weak decays also has a role to play.

External methods are simple in principle. Suppose a state decays into two spin zero particles (for example pions). The angular momentum $J$ of the state must be integral. If there is some axis with respect to which $J_Z = 0$, then relative to that axis the angular distribution of pions (in the centre of mass) will be $[P_J(\cos\theta)]^2$. Try possible axes and look for the highest power of $\cos\theta$ —

there will be some axis in some process giving at least some degree of alignment. Similar but more complex arguments apply to nucleon-pion final states. (See Problem 4.1)

For the hyperons, the strong interactions can produce transversly polarised hyperons ($(\mathbf{p}_1 \times \mathbf{p}_2).\boldsymbol{\sigma}$ is a scalar) and in the weak decays there is a $(1+\alpha\cos\theta)$ dependence of the momentum of the decay pion relative to the hyperon spin. This effect of parity violation is in fact $s$-wave $p$-wave interference and nicely characteristic of spin $1/2$.

Internal evidence is a bit trickier. A nice illustration is provided by $\pi^\circ$ decay: $\pi^\circ \rightarrow \gamma\gamma$. The photons are transverse (remember $\nabla.\mathbf{E} = 0$ becomes $\mathbf{k}.\mathbf{E} = 0$ for plane waves) and the electric field vectors $\boldsymbol{\varepsilon}$ are perpendicular to a relative momentum vector $\mathbf{p}$ in the centre of mass. The final state is described (internally) by $\psi(\mathbf{p}, \boldsymbol{\varepsilon}_1, \boldsymbol{\varepsilon}_2)$. It is observed that the $\varepsilon$ vectors are distributed according to $[\mathbf{p}.(\boldsymbol{\varepsilon}_1 \times \boldsymbol{\varepsilon}_2)]^2$. The quantity $\mathbf{p}.(\boldsymbol{\varepsilon}_1 \times \boldsymbol{\varepsilon}_2)$ is a pseudo-scalar and is the amplitude describing the final state. It changes sign under the parity operation (inversion of spatial coordinate) and is rotationally invariant. The final state is thus $J^P = 0^-$ and so given that electromagnetic processes conserve both angular momentum and parity, the (neutral) pion is pseudoscalar. [The polarisation vectors $\boldsymbol{\varepsilon}_1$, $\boldsymbol{\varepsilon}_2$ have not been directly measured, but the $(e^+e^-)$ planes in the rare decay $\pi^\circ \rightarrow (e^+e^-)(e^+e^-)$ are observable.]

Consider a three particle final state — $K^+ \rightarrow \pi^+\pi^-\pi^+$. The pair $\pi^+\pi^+$ consists of identical bosons and so only exists in a state of even angular momentum. Because wave functions behave as $\sim (kr)^L$, the matrix element will die out as the relative momentum $k$ in the $\pi^+\pi^+$ centre of mass goes to zero, UNLESS $L_{\pi^+\pi^+} = 0$. Similarly, if $L_{\pi^-}$, the angular momentum of $\pi^-$ relative to $\pi^+\pi^+$, is not zero, the matrix element will die as $p_{\pi^-} \rightarrow 0$ in the (three particle) centre of mass. In the decay of the charged kaon, the Dalitz plot is uniformly populated — there are no such holes. Hence $L_{\pi^+\pi^+} = 0$, $L_{\pi^-} = 0$ and $J = 0$. [The Dalitz plot is a 2-dimensional scatter plot of $M_{12}^2$ vs $M_{23}^2$ or kinetic energies $T_1$, $T_2$, $T_3$ along three axes at 120° intervals. An element of the area is proportional to (Lorentz invariant) phase space.]

Parity of the kaon? Decays do not help, because the decay is induced by the weak interactions which violate parity. $K^+ \rightarrow \pi^+\pi^\circ$ (s-wave) and this final state has positive parity. The decay $K^+ \rightarrow \pi^+\pi^-\pi^+$ in an overall s-wave leads to a final state which has negative parity (because there are three pions and the pion has negative parity). This was the puzzle that put Lee and Yang on the right track and led to the discovery of parity violation in the weak interactions. We can get the relative $K\Lambda$ parity however. It was done via hypernuclei: $K^- + {}^4\text{He} \rightarrow {}^4_\Lambda\text{H} + \pi^\circ$. The capture of $K^-$ is from an $s$-state, ${}^4$He and ${}^4_\Lambda$H are both spin zero, so the pion (negative parity) is produced in an $s$-wave. Then the $K\Lambda$ relative parity is negative and it is convenient to choose the $\Lambda$ parity positive and the kaon parity negative.

Bubble chambers filled with liquid hydrogen or deuterium were a magnificent tool in the early days (before $\sim$ 1970) for the study of hadron physics. You could SEE everything!

For example, $\pi^- p \to K^\circ \Lambda^\circ$ (associated production)

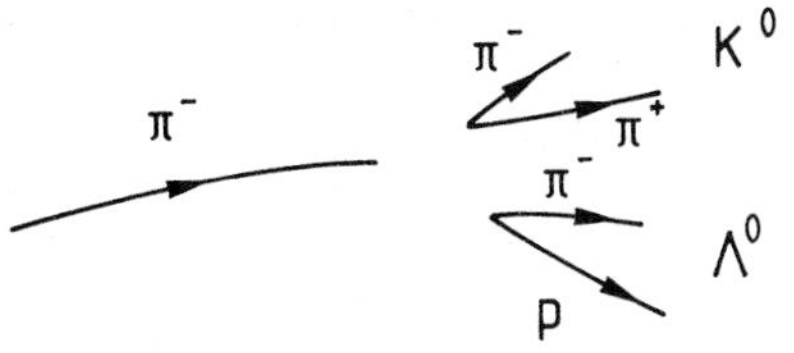


Fig.4.1

Get momenta from curvature in a magnetic field. With incident beams $\sim$ a few GeV/c, velocities are quite low and ionisation $(dE/dx) \sim I_o c^2/v^2$ so the bubble density along the tracks gives a measure of velocity — everything! With cross sections $\sim$ 10mb, you get an interaction every few metres of beam track, and weakly decaying products travel centimetres before decaying and are easily identified. (The spatial resolution is much better than 1mm). Note that an event such as the one sketched above in fig.4.1 is highly over-constrained—if $K^\circ \to \pi^\circ\pi^\circ$ and only the $\Lambda$ is visible, you can reconstruct the missing energy and momentum and find that the recoiling neutral particle is a $K^\circ$ (and not, say, a $\pi^\circ$).

With the complete event and precise measurements, you can also find that in some proportion of the events momentum and energy are missing from the pictures, such that $E^2_{\text{missing}} - p^2_{\text{missing}} \sim 0$—a photon—and the mass recoiling against $K^\circ$ is $\sim$ 1190MeV/c$^2$ ($\Sigma^\circ$) rather than $\sim$ 1115MeV/c$^2$ ($\Lambda^\circ$). The process is $\pi^- p \to \Sigma^\circ K^\circ$ and if you are lucky you will see the photon from the electromagnetic decay $\Sigma^\circ \to \Lambda^\circ\gamma$ convert in the liquid (sometimes). The mass of $\Lambda^\circ\gamma$ can then be constructed directly—$\sim$ 1190 MeV/c$^2$.

Similarly you see

$\pi^- p \to \Sigma^- K^+$ but never $\pi^- p \to \Sigma^+ K^-$

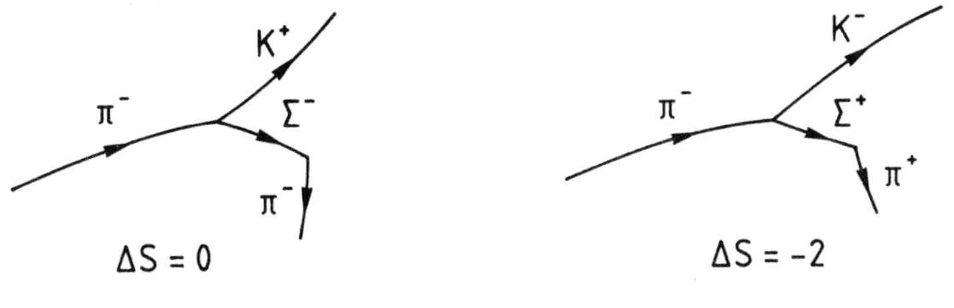


Fig.4.2

You do see $\pi^+ p \to \Sigma^+ K^+$

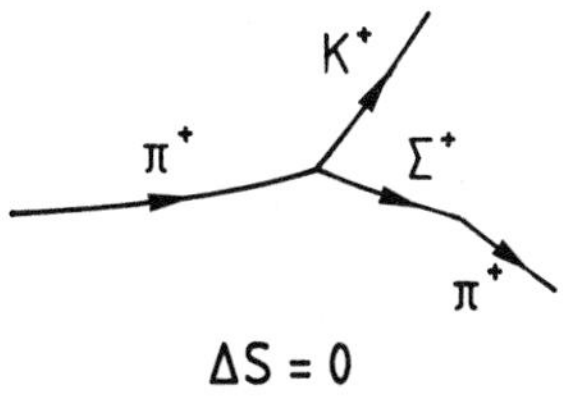


Fig.4.3

If you select negative particles with the kaon mass (using both electric and magnetic fields) from debris resulting from a proton beam impinging on a target, then you see

$$K^- p \to \Lambda^\circ \pi^\circ \qquad K^- p \to \Xi^- K^+$$
$$\Sigma^+ \pi^- \qquad \to \Xi^\circ K^\circ$$
$$\Sigma^- \pi^+ \qquad \text{and so on} \ldots$$

The magnetic moments of the octet baryons have been measured (by rotating the spins in a magnetic field, using the weak decay as an indicator of spin direction; and for $\Sigma^-$ from the fine structure of $\Sigma^-$ atoms). With the SPS and Fermilab accelerators beams of hyperons have been produced, putting time dilation to work...

What about the resonance states, hadrons unstable against the strong interactions? The earliest example was $\Delta^{++}$ which appeared as a resonance in $\pi^+ p$ elastic scattering...

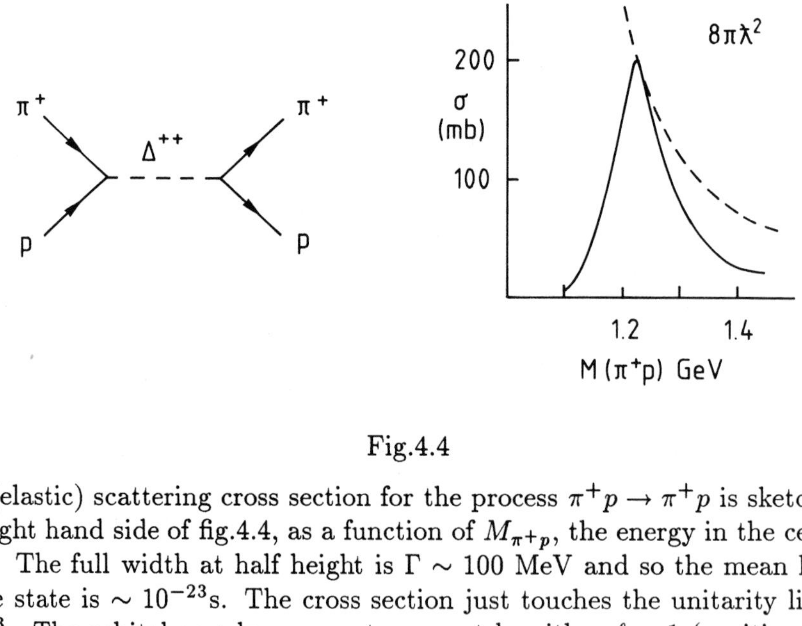


Fig.4.4

The (elastic) scattering cross section for the process $\pi^+ p \to \pi^+ p$ is sketched on the right hand side of fig.4.4, as a function of $M_{\pi^+ p}$, the energy in the centre of mass. The full width at half height is $\Gamma \sim 100$ MeV and so the mean lifetime of the state is $\sim 10^{-23}$s. The cross section just touches the unitarity limit for $J = \frac{3}{2}$. The orbital angular momentum must be either $\ell = 1$ (positive parity) or $\ell = 2$ (negative parity).The differential cross section takes the form

$$\frac{d\sigma}{d\cos\theta} \sim 1 + 3\cos^2\theta$$

which obtains for $\ell = 1$. The $\Delta$ is thus $J^P = \frac{3}{2}^+$.

The other three members of this isospin quartet ($\Delta^+$, $\Delta^\circ$, $\Delta^-$) were found in a similar way. It should be remembered that this is an unusually straightforward case, in that the $T = \frac{3}{2}$ $J^P = \frac{3}{2}^+$ amplitude is overwhelmingly dominant at centre of mass energies $\sim 1200 - 1300$ MeV. At higher energies resonances are denser, broader and highly inelastic and the extraction of the partial wave amplitudes becomes a very tricky business. For example, between centre of mass energies of 1900 and 2000 MeV, there are six $T = \frac{3}{2}$ resonances to which the Particle Data Group attaches either three or four stars, where four star status means "good, clear and unmistakable" and three stars implies a candidate is "good, but in need of clarification or not absolutely certain".

Nonetheless, a state coupled to an accessible two body input channel is best studied this way—FORMATION PROCESSES. The accessible two body input channels are limited. The only targets are $p$ and $n$ (bound in the deuteron) and apart from $p$ and $\bar{p}$ the only particles that can be produced in intense beams of well defined momentum are the charged pions and kaons. The dominant channels are thus

$$\begin{pmatrix} \pi^+ \\ \pi^- \\ K^+ \\ K^- \end{pmatrix} \times \begin{pmatrix} p \\ n \end{pmatrix}$$

It is not possible to make $\Xi(S = -2)$ in a formation process. It has to be produced. Similarly $\Sigma(1385) \to \Lambda\pi$ is below the $Kp$ threshold but appears in the PRODUCTION PROCESS $K^-p \to \Lambda\pi^+\pi^-$ as a resonance in $\Lambda\pi$.

The channels

$$\begin{pmatrix} \pi^+ \\ \pi^- \\ K^- \end{pmatrix} \times \begin{pmatrix} p \\ n \end{pmatrix}$$

are stiff with resonances. The $K^+$ channels present a great contrast. There are no unambiguous resonances in these channels and we take the view that there are no positive strangeness baryons.

A nice example of a production mechanism is illustrated in fig.4.5

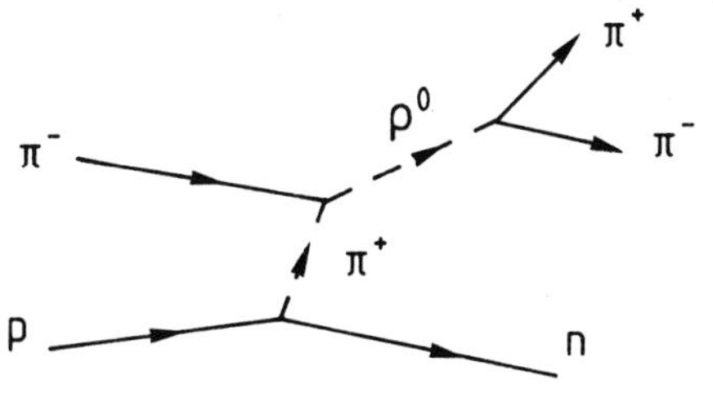


Fig.4.5

Of course you do not see the $\rho$ in a bubble chamber (it only travels $\sim$ 1fm and this one is neutral anyway) but construct the mass distribution of $\pi^+\pi^-$.

This reaction was studied extensively using hydrogen bubble chambers, and with far greater statistics using electronic techniques when beams of $\pi^- \gtrsim 20$ GeV became available. With virtual pion exchange the momentum transfer is small and the two outgoing pions are collimated in the forward direction. Arrays of spark chambers before and after a bending magnet of relatively modest aperture sufficed for reconstruction of the pions. The $\pi^+\pi^-$ mass spectrum peaks at $\sim$ 770 MeV and has a width of $\sim$ 150 MeV; the lifetime is $\sim 10^{-23}$s.

At small momentum transfer the above diagram, fig.4.5 — virtual pion exchange — dominates. We have almost real $\pi^+\pi^-$ elastic scattering. The quantity $\cos\theta = \hat{\mathbf{p}}_{\pi^-}(\text{out}).\hat{\mathbf{p}}_{\pi^-}(\text{in})$ is distributed according to $\cos^2\theta$ (working in the $\pi^+\pi^-$ centre of mass). This is $[P_1(\cos\theta)]^2$ so the angular momentum and parity of the $\rho$ are $J^P = 1^-$—a vector meson. This is an external method of spin-parity determination. (In fact there is also strong $s$ wave scattering.)

The $\omega^\circ$ meson was first seen in

$$\bar{p} + p \rightarrow \omega + X(= \pi^+\pi^-)$$

and appears as a narrow spike in the mass distribution of $\pi^+\pi^-\pi^\circ$. An internal method of getting $J^P$ works here. First, there is no charged $\omega$. Therefore the isospin is zero. The width $\Gamma$ is $\sim$ 10MeV — a bit narrow but certainly a strong decay and in strong interactions isospin is conserved. Since a pion has $T = 1$, every dipion combination must have $T = 1$ to make a $3\pi$ state with $T = 0$. The dipion state with $T = 1$ is odd under interchange (like $J = 1$). Therefore the space part of the $3\pi$ final state, represented through the momentum vectors, must be odd under interchange of a pair to give overall (Bose) symmetry. The momentum vectors carry information about spin and parity. One can write down leading terms (almost) by inspection.

Suppose $0^+$? Then $\psi(\mathbf{p}_1, \mathbf{p}_2, \mathbf{p}_3)$ must be odd under parity and rotationally invariant (odd because there are $3\pi$). The pseudoscalar $\mathbf{p}_1.(\mathbf{p}_2 \times \mathbf{p}_3)$ is odd on swapping 2 and 3 but is also zero because (in the $3\pi$ centre of mass) the three momentum vectors are coplanar. A $T = 0$ state of three pions cannot be $0^+$.

Suppose $0^-$? Then $\psi(\mathbf{p}_1, \mathbf{p}_2, \mathbf{p}_3)$ is a scalar, even under parity, and must be odd under interchange. Something like $\mathbf{p}_1.(\mathbf{p}_2 - \mathbf{p}_3)$? Add two more such terms to get the symmetry right under interchange of any pair? Such a sum vanishes. We can form a product with the right properties:

$$[\mathbf{p}_1.(\mathbf{p}_2 - \mathbf{p}_3)][\mathbf{p}_3.(\mathbf{p}_1 - \mathbf{p}_2)][\mathbf{p}_2.(\mathbf{p}_3 - \mathbf{p}_1)] \tag{4.1}$$

This vanishes strongly as the momentum vectors take on a Mercedes-Benz configuration 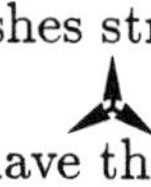 (in the middle of the Dalitz plot) and also vanishes when any pair have the same momentum or the difference is perpendicular to the odd momentum vector; a characteristic pattern, which does not match the $\omega^\circ$ Dalitz plot.

Suppose $1^-$? $\psi(\mathbf{p}_1, \mathbf{p}_2, \mathbf{p}_3)$ must be an axial vector (even under parity). The answer is obvious:

$$\mathbf{p}_1 \times \mathbf{p}_2 + \mathbf{p}_2 \times \mathbf{p}_3 + \mathbf{p}_3 \times \mathbf{p}_1 \tag{4.2}$$

which is proportional to $\mathbf{p}_1 \times \mathbf{p}_2$ because $\mathbf{p}_1 + \mathbf{p}_2 + \mathbf{p}_3 = 0$. This only vanishes when any two are parallel (which is along the rim of the Dalitz plot). This is the $\omega^0$ pattern—$\omega^\circ$ is another $1^-$ (vector) meson.

The above method of constructing appropriate amplitudes by inspection or trial and error becomes progressively less satisfactory as the spin or complexity of the final state increases, but there exist systematic treatments (see for example

C. Zemach *Phys. Rev.* **140B** 97 (1965); R.J. Cashmore *in* Phenomenology of Particles at High Energies, ed. R.L. Crawford and R. Jennings, New York, London 1974). It should also be noted that the overall constraints imposed by spin and parity are modulated by resonances: a three particle final state is frequently reached by decay into another unstable hadron which subsequently decays. For example at three pion masses $\sim 1.3$ GeV there exist

$$a_2(J^P = 2^+) \to \rho\pi \quad D-\text{wave}$$
$$a_1(J^P = 1^+) \to \rho\pi \quad S-\text{wave}$$

where the $\rho$ subsequently decays to $\pi\pi$. There are strong $\rho$ bands in the three pion Dalitz plot.

When a hadron is produced which decays into three light hadrons, stable against decay via strong processes, in general both internal and external information is available—the Dalitz plot population and in addition the orientation of the three particle plane with respect to axes constructed from the beam direction and the normal to the production plane. The subject becomes extremely technical: there was a period in the 1970s when three particle partial wave analysis was a major industry (but it is now becoming a lost art).

Angular momentum analysis of final states containing more than three particles has remained undeveloped, except in those simple cases where some of the particles are the decay products of a relatively narrow resonance, for example

$$b_1 \to \omega^\circ\pi; \qquad \omega^\circ \to \pi^+\pi^\circ\pi^-$$
$$\eta' \to \eta\pi^+\pi^-; \qquad \eta \to \pi^+\pi^\circ\pi^-$$

### 4.3 Hadrons and quarks.

The hadron ground states were listed at the beginning of this chapter: 8+1 $0^-$ mesons and 8 $\frac{1}{2}^+$ baryons. The first excited states consist of $8+1$ $1^-$ mesons and 10 baryons:

| T | S | MESONS | M (MeV) | T | S | BARYONS | M (MeV) |
|---|---|---|---|---|---|---|---|
| $\frac{1}{2}$ | +1 | $K^{*+}$ $K^{*\circ}$ | 890 | $\frac{3}{2}$ | 0 | $\Delta^{++}$ $\Delta^+$ $\Delta^\circ$ $\Delta^-$ | 1232 |
| 1 | 0 | $\rho^+$ $\rho^\circ$ $\rho^-$ | 770 | 1 | −1 | $\Sigma^+$ $\Sigma^\circ$ $\Sigma^-$ | 1385 |
| 0 | | $\omega^\circ, \phi$ | | | | | |
| $\frac{1}{2}$ | −1 | $\bar{K}^{*\circ}$ $K^{*-}$ | (890) | $\frac{1}{2}$ | −2 | $\Xi^\circ$ $\Xi^-$ | 1530 |
| | | | | 0 | −3 | $\Omega^-$ | 1672 |
| | | $8+1$ | | | | 10 | |
| | | $J^P = 1^-$ | | | | $J^P = \frac{3}{2}^+$ | |

These all decay strongly with the exception of $\Omega^-$ which is below threshold for $\Xi(1320)K$. It gets rid of one unit of strangeness via a weak decay, has mean life $\tau = 0.8 \times 10^{-10}$s and leaves a visible track in a bubble chamber. The 1, 8, 10 are SU(3) flavour multiplets and a good guess at the symmetry breaking rules implied for the decuplet equal spacing as a function of increasing strangeness. The existence and mass of $\Omega^-$ were confidently predicted before it was discovered.

The basic multiplet of an SU(3) is a triplet. (The basic multiplet of an SU(2) like isospin is a doublet). These $0^-, 1^-; \frac{1}{2}^+, \frac{3}{2}^+$ multiplets are easily built out of three flavours of spin $1/2$ quarks—*uds*. The rule is simply that mesons are $q\bar{q}$ and baryons are $qqq$. This is a very restrictive rule but almost all known hadronic states satisfy it.

The simplest $0^-$ construct is a $q\bar{q}$ spin singlet with no orbital angular momentum — the negative parity comes from the opposite intrinsic parity of fermion and antifermion, from the Dirac equation. Keep zero orbital angular momentum, but take a spin triplet and 9 vector mesons are obtained. Add one unit of orbital angular momentum and there are again $3 \times 3 = 9$ $q\bar{q}$ flavour combinations, but triplet $L = 1$ states contain $2^+, 1^+, 0^+$ and the singlet $L = 1$ state is another $1^+$ — 4 families of 9 mesons each. All $2^+$ and most of both $1^+$ families are identified present and correct.

The $\frac{3}{2}^+$ decuplet is easily accounted for with three quarks in an overall *s*-wave with spins summing to $J = \frac{3}{2}$. Such states are symmetric under interchange of a pair of identical fermions and this simple model apparently violates the Pauli exclusion principle. It is however hardly credible that low lying states are other than *s*-wave and there is a real problem here. The $\frac{1}{2}^+$ octet is *s*-wave again but with quark spins summing to $\frac{1}{2}$ and these states are also symmetric under interchange of identical fermions. Orbital excitations of baryons are more complicated than those of mesons because there are two ways of introducing an orbital excitation.

With just three kinds of quark we have to associate charges $+\frac{2}{3}, -\frac{1}{3}, -\frac{1}{3}$ with $u, d, s$ respectively. The quarks also have fractional baryon number—but there is no reason for that to worry anyone. With the charges of the quarks identified, we can then infer that the wave functions really are symmetric—from the octet baryon magnetic moments†.

The picture is compelling. The weak decays are all the effects of the fundamental 4-fermion interaction (mediated by $W^{\pm}$), as illustrated in fig.4.6, and we have some insight into the internal machinery mentioned in Ch.3.

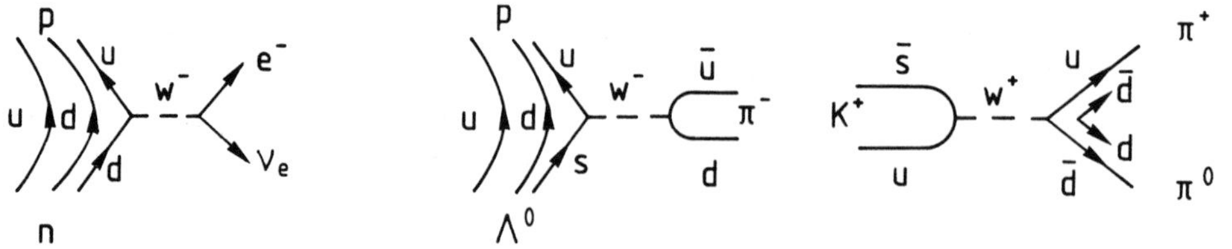


Fig.4.6

The $\Delta^{++}(uuu)$ is formed from $\pi^+(u\bar{d})$ and $p(uud)$ by the annihilation of a $d\bar{d}$ pair, and will live until a $d\bar{d}$ pair is created from the available energy, illustrated in fig.4.7.

† This is discussed in detail in Ch.13.

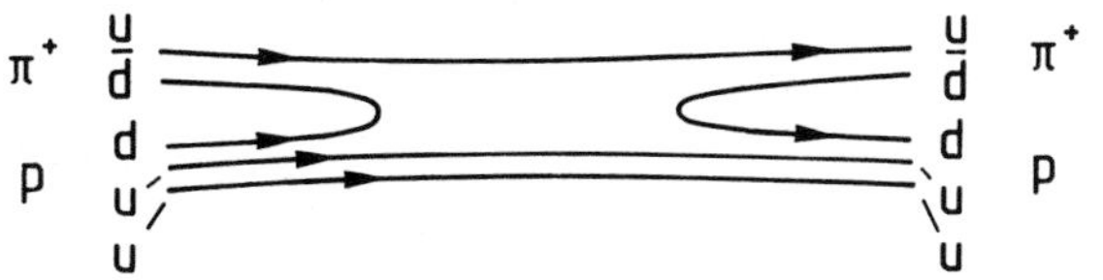


Fig.4.7

Such states appear as well defined resonances. In contrast, the antiquark present in the $K^+p$ channel is $\bar{s}$ and cannot annihilate with a valence quark present in the proton. Thus the process shown in fig.4.8a) exists, but $K^+p$ scattering (fig.4.8b)) cannot involve the formation of a three quark intermediate state

(a) (b)

Fig.4.8

Then $K^+p$ shows no resonance because (baryon) resonances are bound states of three quarks. So far this is only a systematisation rather than an explanation, but in the language of nuclear physics the three quark states are analogous to structure resonances, whereas any resonant effects in $K^+p$ or $K^+n$ would be analogous to shape resonances, scattering from a simple potential hole.

## 4.4 Quark model puzzles.

At first sight the simple quark model, due to Gell-Mann and to Zweig, is completely ridiculous. Its only virtue is that it works. The model in its simple form offers no explanation of the following puzzles.

First, the obvious interpretation of the $\frac{3}{2}^+$ decuplet runs foul of the Pauli exclusion principle. The $J_z = +\frac{3}{2}$ corner states ($uuu$, $ddd$, $sss$) apparently contain three identical quarks in identical quantum states. More generally, the baryon octet and decuplet are symmetric under interchange of any pair of identical quarks, yet the spin $1/2$ quarks should be fermions. Secondly, the requirement that mesons are $q\bar{q}$ states and baryons are $qqq$ states was imposed in order to explain the hadron spectra and is not a result of the model. There is no reason for the absence of mesons with baryon number $\frac{2}{3}$, $qq$ states, and above all the simple model provides no explanation for the absence of free quarks.

All these puzzles are resolved with the introduction of a single new concept — colour.

## Problems

4.1 i) Calculate the relative cross sections for the processes

$$\pi^+ p \to \pi^+ p$$
$$\pi^+ n \to \pi^+ n$$
$$\pi^+ n \to \pi^\circ p$$
$$\pi^- p \to \pi^- p$$
$$\pi^- p \to \pi^\circ n$$
$$\pi^- n \to \pi^- n$$

at a centre of mass energy $\sim$ 1220 MeV, where the $T = \frac{3}{2}$ channel dominates. [See eqs.(2.17)]

ii) Verify the claim that $\pi p$ scattering in the $\frac{3}{2}^+$ $p$-wave gives rise to the angular distribution $1 + 3\cos^2\theta$, where $\theta$ is the $\pi$ scattering angle in the centre of mass. [The isospin couplings for $T = 1 \otimes T = \frac{1}{2}$ are identical to the angular momentum coupling for $\ell = 1 \otimes s = \frac{1}{2}$. Identify the $\ell - s$ composition of $J = \frac{3}{2}$, $\ell = 1$ and check your result by showing that it is an eigenstate of $\mathbf{J}^2 = (\boldsymbol{\ell} + \mathbf{s})^2$ with eigenvalue 15/4.]

4.2 The momenta of particles are obtained from their radii of curvature in magnetic fields. Calculate the accuracy with which momenta may be determined a) by digitising 10 points with an accuracy of 300$\mu$m along 50cm of track in a hydrogen bubble chamber operated in a magnetic field of 10kG (1T) b) with a magnetic spectrometer in which the magnet has a bending power $\int \mathbf{B}.d\boldsymbol{\ell} =$ 10kG-m. The particles are detected by five planes of spark chambers or wire chambers located 2m fore and aft of the magnet. The adjacent planes in each pack are separated by 10cm and the position of the track in each plane is located with an accuracy of 500$\mu$m.

The cross section for the process

$$\pi^- p \to n\pi^+\pi^-$$

is $\sim$ 0.4mb at an incident momentum of 17 GeV/c. The total $\pi^- p$ cross section is $\sim$ 25mb. The pions from each accelerator burst arrive within $\sim$ 1ms for bubble chamber operation, but a slow spill over $\sim$ 1s is used for counter techniques. Consider the limitations on the rates at which $\pi^- p \to n\pi^+\pi^-$ can be accumulated a) in a hydrogen bubble chamber of effective length 1m b) electronically, assuming the apparatus is paralysed for $\sim$ 1ms while recording each event for which it is triggered.

4.3 Parity violation in hyperon decay (for example $\Lambda^\circ \to p\pi^\circ$) is manifested in an asymmetric distribution of the decay products relative to the hyperon spin $\boldsymbol{\sigma}$ of the form

$$1 + \alpha\boldsymbol{\sigma}.\mathbf{p}$$

where $\mathbf{p}$ is a unit vector along the direction of the final state baryon. Hyperons may emerge transversely polarised from production by strong process, the spin being distributed according to

$$1 + \beta\sigma.(\mathbf{p}_1 \times \mathbf{p}_2)$$

where $\mathbf{p}_1 \times \mathbf{p}_2$ is of unit length and perpendicular to the production plane. Explain carefully why the asymmetry demands parity violation in the decay process, whereas the transverse polarisation is produced by strong interactions which do not violate parity.

4.4 When $\pi^-$ are stopped in liquid hydrogen the processes

$$\pi^- p \rightarrow \pi^\circ n; \quad \pi^\circ \rightarrow \gamma\gamma$$
$$\pi^- p \rightarrow \gamma n$$

occur. Calculate the energy of the single $\gamma$ from the second process and the shape of the $\gamma$ energy spectrum arising from the first. [Pay particular attention to the maximum and minimum $\gamma$ energies, and remember that the pion is spin 0.]

4.5 The spin of the $K^*(890)$ was established from the angular distribution of the outgoing kaon relative to a $z$ axis along the direction of the incoming kaon and a $y$ axis normal to the production plane, all directions defined in the $K^+\pi^-$ centre of mass.

In the process

$$K^+ p \rightarrow \Delta^{++} K^{*\circ}$$

the $K^+$ scattering angle is distributed according to $\cos^2\theta$, for small momentum transfer. Consider (i) the expected form of the distribution of the azimuthal angle $\phi$ in the $K^+\pi^-$ rest frame (ii) the expected distribution of the $\pi^+$ in the $\pi^+ p$ rest frame.

In the charge exchange process

$$K^+ n \rightarrow p K^+ \pi^-$$

the $K^+$ scattering angle in the $K^*(890)$ frame is again distributed according to $\cos^2\theta$. What is expected to be the form of the distribution in the $K_2^*(1430)$ rest frame? [The $K_2^*$ is a $2^+$ meson].

In the process $K^+ p \rightarrow K^\circ \pi^+ p$ the kaon scattering angle, in the $K^*(890)$ rest frame, is distributed approximately isotropically and the azimuthal angle according to $\sim 1 + \sin^2\phi$. Reconcile this observation with the properties of $K^*(890)$ revealed in the charge exchange processes.

4.6 The isoscalar vector meson $\phi$ has a mass of 1020 MeV and a width of 4.5 MeV. It decays into

$$\begin{array}{rl} K^+K^- & (50\%) \\ K^\circ \bar{K}^\circ & (34\%) \\ \pi^+\pi^\circ\pi^- & (15\%) \end{array}$$

It is so close to $K\bar{K}$ threshold that phase space and centrifugal barrier effects account for the difference between the kaonic branching ratios. Consider how the marked preference for kaonic decay can be accounted for in terms of the quark model. (The existence of the pionic decay mode is more of a problem.)

# 5. COLOUR, CONFINEMENT AND STRINGS.

## 5.1 Successes of and problems with the quark model.

The low mass hadron multiplets can be built from very simple combinations of three flavours of spin $1/2$ quarks, $uds$. Mesons are built from $q\bar{q}$, yielding $8+1$ multiplets for a given spin-space structure and baryons are built from $qqq$. If it is supposed that the quarks in the $\frac{3}{2}^+$ states are symmetric in spin and space coordinates, it is trivial to show that a decuplet is generated. The requirements that the ground state baryon octet is $s$-wave with total spin $1/2$ and that the spin-space structure is symmetric under exchange of any two identical fermions ensures an octet†. It is nearly true to say that all established hadron states are bound states in the quark sector and only decay strongly to other hadrons through the creation of $q\bar{q}$ pairs from the available energy. The strong interactions among hadrons are predominantly a matter of annihilation and creation of quark pairs. The electromagnetic and weak interactions of hadrons are understood in terms of fields coupling directly to the bound quarks, and processes such as $\Lambda \to p\pi^-$ really are four fermion interactions, mediated by an intermediate $W$ boson field.

There are of course multiquark bound states which are stable—nuclei. The characteristic nucleon binding energy is only $\sim$ 8 MeV/nucleon, whereas the hadronic energy scale is $\sim$ GeV. Nuclei may be regarded as molecular states of quark chemistry. There may be unstable molecular states of the form $(q\bar{q})(q\bar{q})$ or $(q\bar{q})(qqq)$; the former may be important in the $T=0$ and $T=1$ $0^+$ states with a mass of $\sim$ 1 GeV, which have a curious prediliction for decay into $K\bar{K}$. If there are resonances in the $KN$ (as opposed to $\bar{K}N$) system they are undoubtedly molecular.

At this stage the successes of such a quark model are counterbalanced by a number of problems:

(i) Why are there no free quarks? It is the case that high energy interactions produce swarms of (predominantly) pions, not swarms of free quarks.

(ii) If there exist $q\bar{q}$ and $qqq$ bound states, why are there no states with composition $qq$? Such states would have integral spin and baryon number $\frac{2}{3}$.

(iii) Why are there such meagre signs of resonance in states with quark composition more complicated than $q\bar{q}$, $qqq$? Such states as exist (nuclei) or may exist are molecular rather than atomic.

(iv) Why are the baryon states symmetric under exchange of a pair of identical quarks (or more generally symmetric in terms of spin-space-flavour)? The quarks must be fermions.

We may add one more puzzle which has not been mentioned so far.

(v) Nuclear potentials have exponential tails at large distances, not power law tails. If quarks were bound into hadrons through some conventional potential, quark motion in one nucleon would polarise quark motion in another and give rise to long range, power law, dipole-dipole interactions—Van der Waals forces. They do not exist.

---

† The structure is considered in detail in Ch.13.

All these puzzles can be understood in terms of a very simple picture—confined colour.

### 5.2 Colour.

The existence of $s$-wave baryon states $\Delta^{++}(uuu)$, $\Omega^{-}(sss)\ldots$ can be reconciled with the Pauli principle by postulating the existence of a new internal degree of freedom, which must be at least three valued. Spin-space-flavour symmetric three quark states would be permitted because with at least three values of the new degree of freedom the three $u$ quarks in $\Delta^{++}$ could all be distinguishable in principle. Such a new degree of freedom—now called colour—is probably not more than three valued because all protons are identical. This idea, introduced by Greenberg in the very early days of the quark model, is of the kind claimed by philosophers ***ad hoc***, and at the present stage it explains nothing, merely permits the existence of the symmetric baryons.

There is however direct evidence for the existence of this degree of freedom, and the idea of colour is of great value in explaining all the other puzzles we have listed as soon as the possibility of colour having a dynamical role is considered. The hypothesis then ceases to be ***ad hoc***.

The direct evidence is beautifully simple. In $e^+e^-$ interactions lepton pairs are produced by the processes shown in fig.5.1.

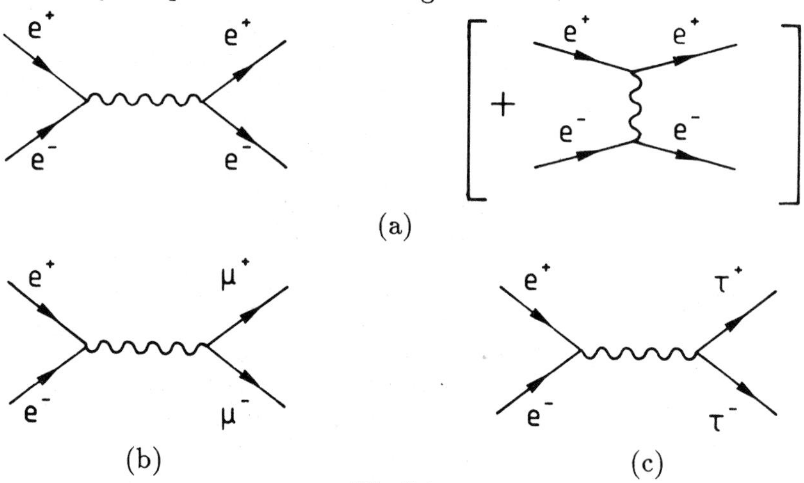


Fig.5.1

For the annihilation diagrams, there is a coupling $e$ at each vertex and the photon propagator is $\sim \frac{1}{s}$, where $s$ is the square of the centre of mass energy (remember that the photon is massless). The amplitude is (using natural units, $\hbar = c = 1$)

$$\mathcal{A} \sim \frac{e^2}{s} \tag{5.1}$$

and because the phase space is $\sim s$, the cross section is

$$\sigma \sim \frac{e^4}{s} \tag{5.2}$$

This much should be clear from the elementary discussion of propagators given in Ch.2 and (5.2) is necessary from dimensions alone. A complete calculation (which will be presented in Ch.9) yields

$$\sigma = \frac{4\pi}{3}\frac{e^4}{s} \quad ; \quad \frac{d\sigma}{d\Omega} = \frac{e^4}{4s}(1+\cos^2\theta) \tag{5.3}$$

where $\theta$ is the angle between the incident $e^+$ and the outgoing $\mu^+$, in the centre of mass. The data agree.

Above centre of mass energies $\sqrt{s} \sim$ a few GeV the process $e^+e^- \to$ hadrons dominates the total annihilation cross section. Furthermore, at high energies $\sim$ 30 GeV (PETRA and PEP) the hadrons appear for the most part in two well collimated jets, back to back. The jet axis is distributed relative to the beams as $(1+\cos^2\theta)$; just like muons. We therefore suppose that hadron production takes place through a primordial process $e^+e^- \to q\bar{q}$, followed by materialisation of the hadrons (fragmentation) through some as yet unspecified process. The angular distribution of the jets presumably reflects that of the quarks and is consistent with spin $1/2$ quarks. The cross section is then

$$\sigma_{e^+e^-\to hadrons} = \frac{4\pi}{3}\frac{e^2}{s}\sum e_q^2 \tag{5.4}$$

and writing $e_q = ef_q$

$$R = \frac{\sigma_{e^+e^-\to hadrons}}{\sigma_{e^+e^-\to\mu^+\mu^-}} = \sum f_q^2 \tag{5.5}$$

Below $\sim$ 3 GeV ($\sim c\bar{c}$ threshold) $uds$ contribute to the sum

$$\left(\frac{2}{3}\right)^2 + \left(\frac{1}{3}\right)^2 + \left(\frac{1}{3}\right)^2 = \frac{2}{3}$$

Above $\sim$ 10 GeV ($\sim b\bar{b}$ threshold) $udscb$ contribute

$$\left(\frac{2}{3}\right)^2 + \left(\frac{1}{3}\right)^2 + \left(\frac{1}{3}\right)^2 + \left(\frac{2}{3}\right)^2 + \left(\frac{1}{3}\right)^2 = \frac{11}{9}$$

Experimentally, $R \sim 2$ for $\sqrt{s} < 3$ GeV and $R \sim 4$ from just above $\sqrt{s} = 10$ GeV to the highest energies at which measurements have been made, $\sim$ 50 GeV. Within the framework of this model for hadron production, the additional factor of three is compelling evidence for an additional three valued quark degree of freedom. Instead of making $u\bar{u}$ the process makes red, blue and green $u\bar{u}$ pairs, and the cross section is calculated under the assumption that the colours may be treated as distinguishable. The incontrovertible points are first, that $R$ is constant between thresholds and secondly that $R$ is three times as large as a calculation without colour gives.

### 5.3 Dynamical colour.

Let us now see how colour can take on a dynamical role as opposed to being a mere counting device. In the decuplet of $3/2^+$ baryons the spin-space wave function of the quarks is totally symmetric under interchange of any pair. The obvious example is $\Delta^{++}(+3/2)\ u\uparrow u\uparrow u\uparrow$. To retain the Pauli principle in the form that fermion wave functions should be totally antisymmetric, we require a totally antisymmetric COLOUR function made from $r$, $b$, $g$. There is just one such function for three quarks, three colours. It is the determinant

$$\frac{1}{\sqrt{6}}\begin{vmatrix} r & b & g \\ r & b & g \\ r & b & g \end{vmatrix} \tag{5.6}$$

Because this is unique, it is a colour singlet. Now suppose that the hadrons are all colour singlets, that is, the colour degree of freedom is attached to quarks 1, 2, 3 according to $\begin{vmatrix} r & b & g \\ r & b & g \\ r & b & g \end{vmatrix}$ regardless of spin, space or flavour. In the meson sector the singlet colour function is

$$\frac{1}{\sqrt{3}}(r\bar{r}+b\bar{b}+g\bar{g}) \tag{5.7}$$

The symmetry of the baryon spin-space wave functions is then explained and we would welcome the existence of some colour exchange mechanism to give physical reality to the colour states, just as charge exchange introduced physical reality into treating a neutron-proton state as $\frac{1}{\sqrt{2}}(pn \pm np)$. We have the opportunity of understanding why it is that we only ever see $q\bar{q}$, $qqq$ states (or very weakly bound clusters of such constituents). If the colours are actually charges for a new field (the colour field of QCD) then a pattern of colour exchange interactions can easily yield colour singlet states ($q\bar{q}$, $qqq$) lying much lower in energy than the coloured states $q$ (triplet of colour), $qq(6+\bar{3})$, $qq\bar{q}$ ... Colour now takes on a dynamical significance†.

It would be very nice if we could keep the masses of colour singlet states finite and drive all other colour configurations to infinite mass. The only physical hadrons would then be colour singlets. This is easily achieved in a simple model—but not in the context of conventional potentials. It seems that the fields must be both self-interacting and strongly coupled. Suppose that colour currents are conserved and that colour exchange forces exist (fig.5.2)

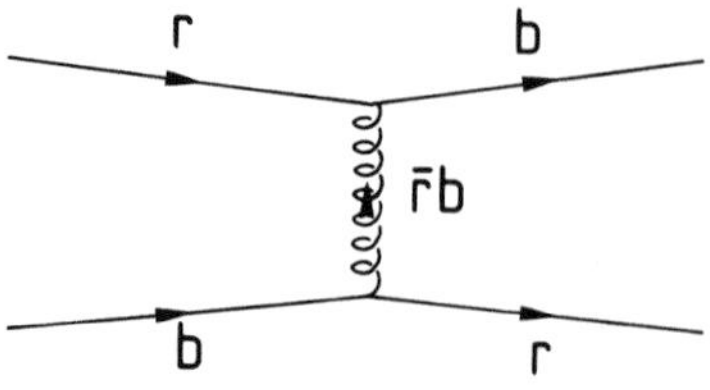


Fig.5.2

Fig.5.2 represents the exchange of a single virtual quantum of the colour field, a coloured gluon carrying colour charges $r\bar{b}$. Since the colour charges couple to

† The calculations are to be found in Ch.12

colour fields, the gluons are self-interacting (and in a strongly coupled theory there may exist glueballs, colourless hadrons composed of glue). There are nine colour combinations, of which one may be taken to be the singlet $\frac{1}{\sqrt{3}}(r\bar{r}+b\bar{b}+g\bar{g})$. Experimentally there is no colour singlet gluon and the gluons form an octet of $SU(3)_{colour}$ [$3 \otimes \bar{3} = 8 \oplus 1$ and there is no 1]. It is easy to write down a theory containing coloured quarks and an octet of coloured (vector) gluons which are massless, quantum chromodynamics or QCD. It has not been solved other than perturbatively, and hadron structure is a thoroughly non-perturbative problem. We have to proceed intuitively, keeping one eye on the data.

## 5.4 Confinement and strings.

If the colour fields $\mathbf{\Sigma}$ satisfy a Gauss' law $\nabla.\mathbf{\Sigma} = 0$ except at a source or sink (colour charge) then the field lines start and terminate on colours. If the fields squeeze themselves into localised flux tubes of colour, the limit of which is a string

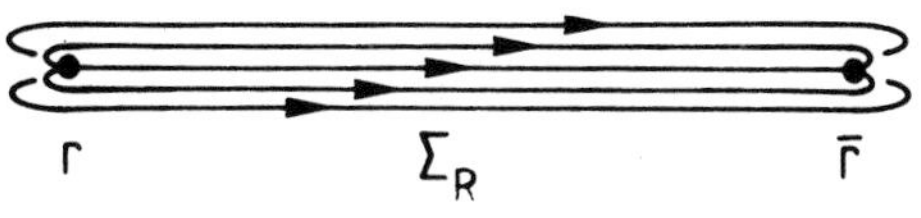


Fig.5.3

then for distances $\gtrsim$ 1fm the field profile transverse to the tube is constant and the energy per unit length $\sim \int \Sigma^2 dA$ is constant. The energy stored in the colour string grows linearly with its length. A string which is not terminated at both ends will contain infinite energy. In such a scheme, only colourless configurations will have finite mass, contributed by the masses and kinetic energies of the quarks and by the length of string joining them. The quarks can never be pulled apart and freed, because you cannot supply infinite energy. If you try, quantum fluctuations create virtual $q\bar{q}$ pairs in the string and as soon as the string is long enough (sufficient energy stored) such virtual quarks tunnel into the real state, sealing off two new ends and making more hadrons (fig.5.4)

Fig.5.4

This picture at once disposes of long range (power law) Van der Waals forces between nucleons. A conventional field spreads out through space, and if the colour fields were like this, colour polarisation of one nucleon would appear in proximity to another. In a string or flux tube picture the fields are confined within tiny tubes and at this level the quarks within one hadron will not feel the colour fields due to quarks in another.

This hypothesis that the colour field between quark and anti-quark takes on a string-like character at separations $\gtrsim$ 1fm is unproved in QCD, although there exist theoretical indications that colour fields behave this way. The second marked regularity of hadron spectroscopy, mentioned at the beginning of Ch.3, not only supports the string picture but yields an important string parameter.

The energy stored in unit length of string is $\sim 0.9$ GeV fm$^{-1}$, or in macroscopic units, 14 tonnes weight: we get this from the other marked regularity of hadron spectroscopy. Classes of hadrons which can be interpreted as a rotational band formed by spinning up some underlying quark configuration exhibit a linear relation between angular momentum and the square of the mass. The slope of these Regge trajectories in the time-like (energy-like) region is $0.9\hbar$ GeV$^{-2}$. This slope is independent of the (light) quark composition, the quantum numbers of the ground state, and whether the band consists of mesons or baryons. [The evidence is in fact not that good—there are no more than five points at most on any leading trajectory if we demand well identified hadron states.]

Consider (massless) $q\bar{q}$ linked by a rotating string and moving at the speed of light

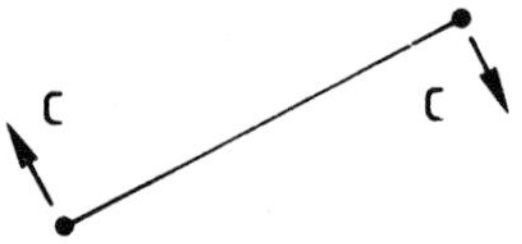


Fig.5.5

At rest the string stores energy $\kappa$ in unit length, and we assume no transverse oscillations on the string (that would correspond to exciting gluonic degrees of freedom). This configuration has the MAXIMUM angular momentum for a given mass, and all of both reside in the string—the quarks have none.

Consider one little bit of string a distance $r$ from the middle (fig.5.6), with the quarks located at fixed distances $R$, and calculate both mass $M$ and angular momentum $J$:

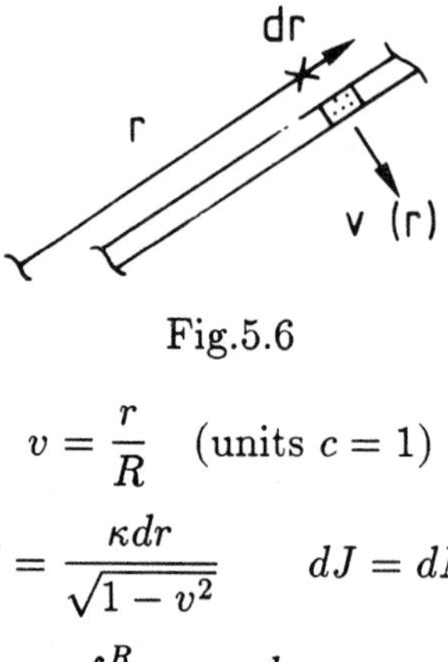


Fig.5.6

$$v = \frac{r}{R} \quad \text{(units } c = 1) \tag{5.8}$$

$$dM = \frac{\kappa dr}{\sqrt{1-v^2}} \qquad dJ = dMvr \tag{5.9}$$

$$M = 2\int_0^R \frac{\kappa dr}{\sqrt{1-(^r/R)^2}} = \pi\kappa R \tag{5.10}$$

$$J = 2\int_0^R \frac{\kappa rv\, dr}{\sqrt{1-v^2}} = \frac{\pi}{2}\kappa R^2 \tag{5.11}$$

Eliminating $R$ we obtain

$$J = \frac{M^2}{2\pi\kappa} \tag{5.12}$$

$(2\pi\kappa)^{-1}$ is the slope of the trajectory and experimentally has the value $0.9\hbar$ GeV$^{-2}$ $\rightarrow 0.9$ GeV$^{-2}$ (natural units). Then $\kappa = 0.18$ GeV$^2$ $= 0.9$ GeV fm$^{-1}$ $=$ 14 tonnes weight.

The field energy between two quarks $\sim$ 1fm apart is $\sim$ 1 GeV. [For comparison, the potential energy of two electron charges 1fm apart is $(-)\frac{1}{137}\times(1/1\text{fm} = 5\,\text{GeV}^{-1}) = 1.4$ MeV and the force is $1/137 \times {}^1\!/5^2 = 2.9 \times 10^{-4}\,\text{GeV}^2 = 23$kg weight.]

We can even make an estimate of the magnitude of the long range colour charge (which must be flavour independent).

Cut a flux tube extending between a colour and an anticolour with a plane—the colour flux through the plane will be $\sim \Sigma A$ where $A$ is the cross sectional area of the tube. Then from

$$\nabla.\Sigma = 4\pi\rho_{colour} \tag{5.13}$$

$$4\pi q = \Sigma A \quad ; \quad \Sigma = 4\pi q/A \tag{5.14}$$

$$\kappa(\text{Energy/unit length}) \simeq \frac{1}{8\pi}\Sigma^2 A = 2\pi q^2/A \tag{5.15}$$

and

$$\kappa = \frac{2\pi q^2}{A}$$

If $A = (1\text{fm})^2 = (5\,\text{GeV}^{-1})^2$ then in natural units

$$q^2 = \frac{\kappa A}{2\pi} = \frac{0.18\,\text{GeV}^2 \times 25\,\text{GeV}^{-2}}{2\pi} = 0.7 \tag{5.16}$$

This is only very rough and is sensitive to the assumed area. It is however indicative that for these long range phenomena $q^2 \simeq \alpha_s \approx 1$ whereas $e^2 = \alpha = 1/137$. Certainly the fields are strongly coupled.

For long range phenomena perturbation theory is unlikely to be much use. In any event, formation of a flux tube is an inherently non-perturbative phenomenon, and is non-linear because of the self interaction of the colour fields.

The string structure of the baryons is not quite so obvious and has a number of interesting features. If the singlet 3-colour combination is also to have FINITE mass, then as far as colour is concerned, $(bg) = \bar{r}$ and $B + G = -R$

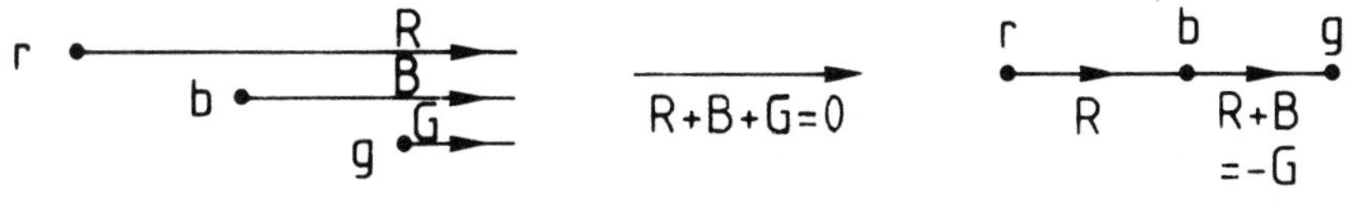


Fig.5.7

It takes two quarks $(bg)$ to sink the Red field produced by a single $(r)$ quark: in colour singlets the $qq$ interaction is half the $q\bar{q}$ interaction†.

The one dimensional picture shown in fig.5.7 also illustrates clearly the origin of the whimsical term colour for the three valued internal degree of freedom attached to the quarks. The superposition of $R$, $B$ and $G$ fields gives nothing.

† See Ch.12.

The superposition of Red, Blue and Green lights gives the sensation of white light—colourless.

There are two possible configurations for spinning strings with three massless quarks:

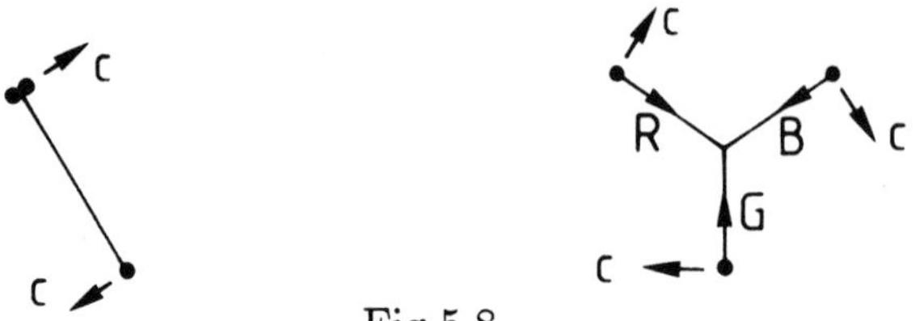


Fig.5.8

The first (quark-diquark) configuration gives the same slope as for the mesons. The second configuration gives $J = \frac{M^2}{3\pi\kappa}$, $^2/_3$ the meson slope $-0.6$ GeV$^{-2}$ rather than 0.9 GeV$^{-2}$. This simple argument favours the first configuration.

### 5.5 Strings and fragmentation.

The string picture gives us a simple way of visualising the confinement of colour and the consequent restriction of bound states of quarks to $q\bar{q}$, $qqq$. It provides us with a semi-classical way of understanding the linearly rising Regge trajectories and the absence of power law nucleon-nucleon potentials. It also provides a semi-classical model of the fragmentation process in which rapidly separating colour charges evolve into jets of colourless hadrons. There exists no theory of the fragmentation process, but in the string model $e^+e^-$ annihilation into hadrons takes place in the following way. First, $e^+e^-$ annihilate into a virtual photon, carrying energy but no momentum in the centre of mass. The virtual photon creates a $q\bar{q}$ pair in precisely the same way as lepton pairs are created. The quarks fly apart, with equal and opposite momenta in the centre of mass. As they fly apart, a string develops between them and as the string stretches it stores energy, approximately 1 GeV fm$^{-1}$. The energy is drawn from the kinetic energy of the quarks, which also lose momentum but continue to travel at the speed of light, in the approximation where light quarks are massless. The process of $e^+e^-$ annihilation has produced a highly excited hadron, consisting of a quark and an antiquark connected by a string of colour. This highly excited hadron is colourless and subsequently decays into colourless lighter hadrons. The amplitude for producing a particular quark flavour is (see (5.1))

$$\mathcal{A} \sim e\frac{e_q}{s}(r\bar{r} + b\bar{b} + g\bar{g}) \tag{5.17}$$

It is not proper to treat $r\bar{r}$, $b\bar{b}$, $g\bar{g}$ pairs as distinguishable, because the physical $q\bar{q}$ state is a colour singlet. However, $(r\bar{r} + b\bar{b} + g\bar{g})$ is $\sqrt{3}$ times a normalised colour singlet $q\bar{q}$ state and so the amplitude for $e^+e^- \to q\bar{q}$ (colour singlet) is

$$\mathcal{A}_{q\bar{q}} \sim \frac{ee_q}{s}\sqrt{3}$$

and the cross section for $e^+e^- \to$ hadrons is

$$\sigma = \frac{4\pi}{3}N_c\frac{e^4}{s}\sum f_q^2 \tag{5.18}$$

where $N_c = 3$, the number of colours, and the sum is taken only over quark flavours. The ratio $R$ (5.5) is again three times that for colourless quarks, but now the factor three appears in the square of the matrix element rather than in the sum over final states.

The second stage is the decay of the highly excited hadron. The energy stored in the string is converted into the energy of the hadrons produced as the string breaks up by the creation of $q\bar{q}$ pairs. These act as new sources and sinks for the colour field and thus cut the original string into little pieces, fig.5.9. The process continues until stable hadrons are reached.

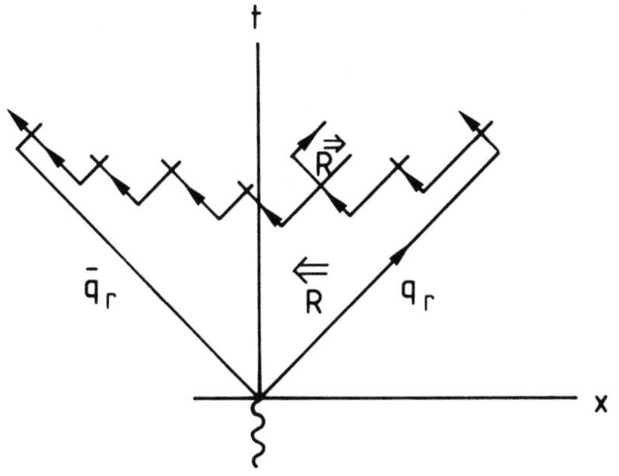


Fig.5.9

In the semi-classical model of string fragmentation, first introduced by Artru and Mennessier, there are two crucial ingredients. The first ingredient is the classical equations of motion of massless quarks attached to a string of constant tension (see Problem 5.3). The second is the assumption that the probability of cutting the string is independent of the space-time coordinates of the cut, provided only that the string exists. This is analogous to the assumption in the semi-classical theory of radioactivity that the probability of a nucleus decaying is independent of time, provided only that the nucleus still exists.

If the quark pairs are produced with small momentum transverse to the string, the hadrons fall into two jets following the directions of the original quarks. The jet axis is therefore distributed according to $1 + \cos^2\theta$ for spin $1/2$ quarks, as observed. An example of a two jet event is shown in fig.5.11.

The vertices at which the string is cut (fig.5.9) are distributed around an hyperbola in $x - t$, and on average slow particles are produced relatively close to the electromagnetic creation point, fast particles progressively further away and at later times. Hadrons close in momentum are produced relatively close together in the rest frame of those hadrons. This property of the string model is confirmed by experiment; $\pi^+\pi^+$ and $\pi^-\pi^-$ pairs with relative centre of mass momentum $\lesssim 100$ MeV are correlated in space and the correlation length is $\sim$ 0.8fm. [This has been established through intensity interferometry with identical bosons, a phenomenon analogous to the Hanbury-Brown-Twiss effect in optical intensity interferometry.]

When electrons, muons and neutrinos are scattered from nucleons, either free or bound in a nucleus, the characteristics of deep inelastic scattering (in which the hadronic final state has mass greater than or of the order of a few GeV) are just those of lepton scattering from free pointlike constituents within the nucleon. These partons are in fact quarks, and this extraordinary result is also easily understood in terms of the picture of hadronic states as quarks connected by a string of colour. A string stretches between the struck quark and the remaining $qq$ system, which is a $\bar{3}$ of colour, and the string breaks up. The picture is like that of fig.5.9, boosted to a frame where one of the original colour charges has low momentum. Fast particles are therefore produced, in the string model, at large distances from the original interaction point—outside the nucleus in the case of a nuclear target. There is evidence that this is so, for the spectrum of fast hadrons produced from a heavy nucleus matches that from deuterium, and this would not be the case if they had to plough through several fermis of nuclear matter.

The colour fields take on a string configuration at separations of $\gtrsim$ 1fm; the string is presumed to be the long range manifestation of QCD. An accelerated quark current may be expected to radiate gluons and if the gluon momentum exceeds a few GeV the radiated gluon probes the current at short distances. In such events the string forms not between, say, $r$ and $\bar{r}$, but in two segments, between a red quark and a gluon carrying colour $\bar{r}b$, and between the gluon and an anti-blue antiquark. The gluon appears at large distances as a kink on the string, moving at the speed of light. The two string segments break up along the line of the string, in the string centre of mass, and three jets result (fig.5.11). Because of the motion of the strings, particles are swept away from the region between the two quark jets. This effect was predicted by a group at Lund before it was first observed by the JADE collaboration, working at PETRA. In figure 5.10 the string configurations are shown as a function of time for $e^+e^-$ annihilation leading to $q\bar{q}g$ and $q\bar{q}\gamma$ at the parton level.

The representation of the long range effects of QCD in terms of flux tubes of colour is not a theory and is a well developed model only at the semi-classical level. It nonetheless has great conceptual advantages and such conclusions as we can tentatively draw from the model are in astonishing accord with the data. The true physics must be rather like this.

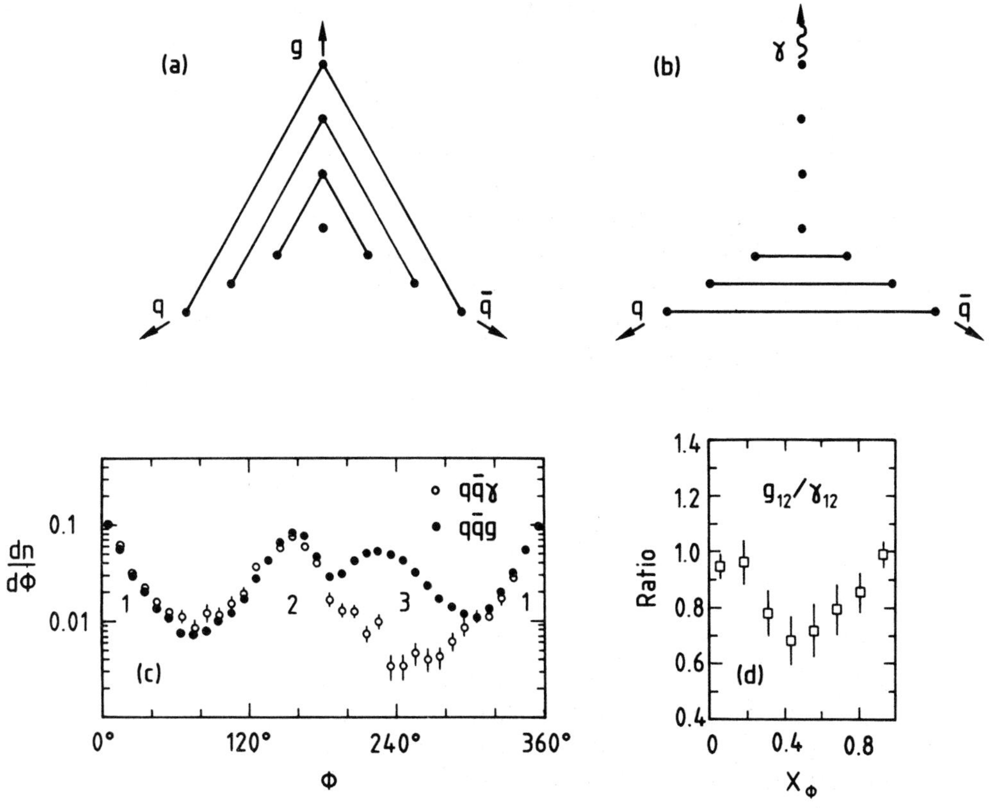


Fig.5.10

(a) and (b) show string configurations, separated by three equal intervals of time, for $q\bar{q}g$ and $q\bar{q}\gamma$ events in which quark and antiquark each carry one third of the available energy. (c) and (d) show data from the Mark II Collaboration at SLAC [*Phys. Rev. Lett.* **57** 1398 (1986)]. The angular distributions of charged particles/event are shown in (c), where $\phi$ is the azimuthal angle in the plane of the event, measured from the axis of the most energetic jet. Note that for $q\bar{q}\gamma$ jet 3 is absent and that the population between the $q$ and $\bar{q}$ jets 1 and 2 is relatively depleted for $q\bar{q}g$. In (d) the ratio is shown as a function of the fraction of the angle between jets 1 and 2.

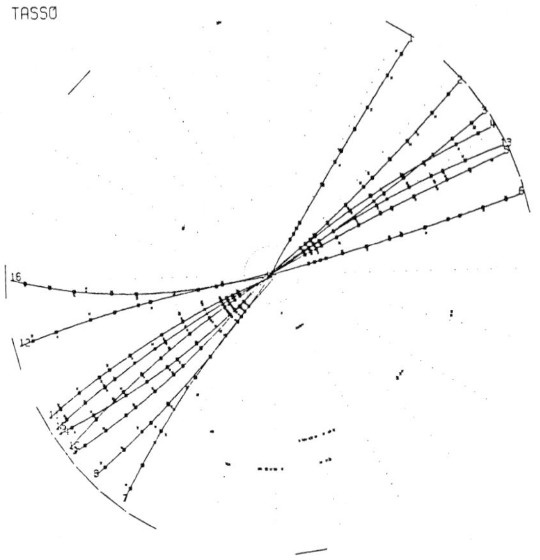


(a)

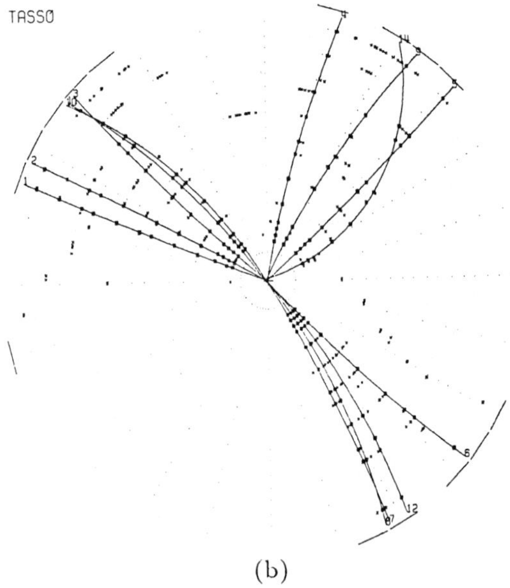


(b)

Fig.5.11

Events classified as 2 jet (a) and 3 jet (b) as observed in the central detector of the TASSO Collaboration. The $e^+e^-$ centre of mass energy was 35 GeV. The dots represent ionisation recorded by wires running parallel to the colliding beams; those hits lying on reconstructed tracks (solid curves) are emphasised. Cylindrical detectors in which charged particles are reconstructed through collection of ions produced by passage through gas are to be found at the core of most apparatus at colliding beam stations.

## 5.6 Chromomagnetism.

In the quark model with three colours the $0^-$ and $1^-$ meson octets are both interpreted as $s$-waves, singlet and triplet in spin. The $\frac{1}{2}^+$ baryon octet and $\frac{3}{2}^+$ baryon decuplet are both overall $s$-wave states. The $1^-$ mesons are split from the $0^-$ by $\sim$ 500 MeV and the decuplet from the baryon octet by $\sim$ 300 MeV. These splittings must be short range effects of the interactions among the quarks, hyperfine (spin-spin) interactions. This will be worked out quantitatively in Chs.12 and 13, but we may anticipate that at short distances the colour field is vector, like electromagnetism, and quarks have chromomagnetic moments. The spin of the gluon should be 1. Three jet events produced in $e^+e^-$ annihilation can be interpreted in terms of the underlying parton momenta, and the $q\bar{q}g$ Dalitz plot (or an equivalent) constructed. It is indeed consistent with massless spin 1 gluons at the parton level†.

## 5.7 References

The physics of the fragmentation process has been largely ignored in the many books on particle physics which have appeared in the last few years. Here are some references to the higher level literature.

An extremely influential paper, couched in elementary terms but containing a core that still survives is R.D. Field and R.P. Feynman, *Nuclear Physics B* **136** 1 (1978).

The string model of fragmentation was first introduced by X. Artru and G. Mennessier, *Nuclear Physics B* **70** 93 (1974).

The modern version of the string model has been extensively reviewed by B. Andersson, G. Gustafson, G. Ingelman and T. Sjöstrand *Physics Reports* **97** 33 (1983). See also X. Artru and M.G. Bowler *Zeitschrift für Physik C—Particles and Fields* **37** 293 (1988)

A review of $e^+e^-$ physics at PETRA is given by Sau Lan Wu *Physics Reports* **107** 59 (1984)

### Problems

5.1 (i) Verify the statement that the Mercedes-Benz configuration for a string model of baryons gives rise to a leading Regge trajectory with slope 2/3 that of the meson trajectories.

(ii) In the string model a glueball is represented by a closed loop of string. A glueball composed of two gluons could be visualised as a straight loop of string with 180° kinks at each end. Find the slope of the leading Regge trajectory for such glueballs and estimate the mass for orbital angular momentum 2.

† You may be bothered that the colour fields are vector, like electromagnetism, and yet both $q\bar{q}$ and $qqq$ colour singlets are bound. It is the existence of colour exchange interactions which makes all the difference (Ch.12).

5.2 You can have a great deal of fun with the string model and relativity. Consider a rigidly rotating string terminated by massless quarks which move at the speed of light. The string is moving with velocity $v_0$ normal to the plane of rotation. Calculate the total energy, the angular momentum and the period directly; the total energy is of course $M\gamma$ and the period is $\tau\gamma$, where $M$ and $\tau$ obtain in the rest frame and $\gamma = (1-v_0^2)^{-\frac{1}{2}}$. [Remember that in (5.9)–(5.11) the energy stored in unit length of string was set equal to $\kappa(1 - v^2)^{-\frac{1}{2}}$, where $v$ is the velocity perpendicular to the string. The above calculation shows that this is not only plausible, but necessary.]

5.3 In two dimensions $(x, t)$ a string can only vibrate longitudinally. The equations of motion of the terminating quarks are

$$\frac{dE}{dx} = \pm\kappa \quad ; \quad \frac{dp}{dt} = \pm\kappa$$

The first is obvious from the definition of $\kappa$ as the energy stored in unit length. The second is not quite so obvious. Show that this pair of equations is covariant (same form, same numerical content) in two dimensions. [This is true for massive terminating quarks as well as for massless quarks.] Draw the $x-t$ trajectories of the ends of the string for the case of massless quarks and for massive quarks, in the centre of mass of the string. For the massless case, construct the trajectories of the end points for half an oscillation in a frame where the string has total momentum $p$, energy $E$. Show that in terms of the light cone variables

$$y = \frac{t+x}{\sqrt{2}} \qquad z = \frac{t-x}{\sqrt{2}}$$

the mass of the string is given by

$$M^2 = 2\kappa^2 \Delta y \Delta z$$

and the rapidity $r$, defined as

$$r = \frac{1}{2}\ln\left\{\frac{E+p}{E-p}\right\}$$

is given by

$$r = \frac{1}{2}\ln\left(\frac{\Delta y}{\Delta z}\right)$$

where $\Delta y$ and $\Delta z$ are the lengths of the light cone segments traced by the massless quarks.

5.4* Consider a straight string vibrating in a direction orthogonal to its motion. The ends are terminated by massless quarks and it is plausible that they move with speed $c$ ($c = 1$ in natural units) in any frame. It is fairly obvious that one equation of motion of the quarks is

$$\frac{dE}{dx} = \pm\frac{\kappa}{\sqrt{1-v^2}}$$

where $v$ is the velocity of translation of the string normal to $x$. It is less obvious that the other equation of motion is

$$\frac{dp}{dt} = \pm\kappa\sqrt{1-v^2}$$

Show that this is required by the principle of Lorentz covariance. [It can be done quite simply, or by writing a general equation

$$\frac{dp_\mu}{d\tau} = F_\mu$$

where $\tau$ is the quark proper time and both $p_\mu$ and $F_\mu$ are four-vectors.] When you have done this you will understand why massless quarks at the end of a rotating string carry neither energy nor momentum. You now have enough information to tackle the problem of two massive quarks connected by a rigidly rotating string. Extract the properties of the leading Regge trajectory for equal quark masses of 0.3 GeV ($s\bar{s}$), 1.5 GeV ($c\bar{c}$) and 5 GeV ($b\bar{b}$).

5.5 The iterative (or recursive) principle of jet evolution was popularised by Field and Feynman. It is supposed that a cascade of hadrons is produced from a primary rapidly moving quark in the following way. A $q\bar{q}$ pair is produced and the $\bar{q}$ combines with the primary quark to form a meson which carries a fraction $(1-\eta)$ of the momentum of the original quark. The remaining fraction $\eta$ is taken by the bereaved $q$, which now acts as the primary quark in the next stage...

If $f(\eta)d\eta$ is the probability that a meson leaves fractional momentum between $\eta$ and $\eta+d\eta$ for the remaining hadrons, and $F(z)dz$ is the probability of finding any (first generation) meson with a fraction of the momentum of the original quark between $z$ and $z+dz$, then $F(z)$ satisfies the integral equation

$$F(z) = f(1-z) + \int_z^1 f(\eta)F(z/\eta)d\eta/\eta$$

Prove it. Solve this equation for the particularly simple case $f(\eta) = 1$. Show that for small $z$

$$F(z) \sim \frac{1}{z}$$

regardless of the form of the function $f(\eta)$.

[The implication is that the multiplicity of a single jet grows logarithmically with jet energy.]

5.6* In the Artru-Mennessier model of string fragmentation, the probability of cutting an existing string in a space-time area $dx\,dt$ at $(x,t)$ is $\mathcal{P}dx\,dt$. As the produced $q\bar{q}$ pair separates, the quarks gaining energy from the string, the string is destroyed in the future light cone of the cut. Show that (i) The probability of a cut occuring at light cone coordinates $(y,z)$ is $\mathcal{P}e^{-\mathcal{P}yz}dy\,dz$. (ii) The mean area in the past of a cut is $1/\mathcal{P}$ and so the cuts are scattered about an hyperbola $\mathcal{P}yz = 1$. (iii) That the cuts are uniformly distributed in the variable $\phi = \frac{1}{2}\ln(y/z)$ [show that $dy\,dz = dT\,d\phi$ where $T = yz$].

Consider two adjacent cuts at $(y_1, z_1)$, $(y_2, z_2)$; $y_2 > y_1$ and $z_1 > z_2$. A first generation secondary string is formed at $(y_2, z_1)$, with mass squared equal to $2\kappa^2\Delta y\Delta z$ and rapidity $\frac{1}{2}\ln(\Delta y/\Delta z)$. Show that the mass squared and rapidity are distributed according to

$$\frac{d^2N}{dm^2dr} = \frac{\mathcal{P}}{2\kappa^2}E_1\left(\mathcal{P}\frac{m^2}{2\kappa^2}\right) \quad \left[E_1(x) = \int_x^\infty \frac{e^{-t}}{t}dt\right]$$

(neglecting end effects). Note that their distribution is uniform in rapidity; string fragmentation is in fact iterative in the sense of problem 5.5.

Convince yourself that high positive rapidity is correlated with large $y_2$, small $z_1$ and high negative rapidity with large $z_1$, small $y_2$.

Show that the fragmentation function of mesons containing a primary quark is given by

$$f(m^2, z) = \frac{\mathcal{P}}{2\kappa^2}\frac{1}{z}\exp\left\{-\frac{\mathcal{P}m^2}{2\kappa^2}\frac{1}{z}\right\}$$

where $\ln(z) = r - r_{\mathbf{max}}$. [Assume throughout that quarks are massless.]

## 6. LEPTON PROBES.

### 6.1 Partons and probes.

The extended charge structure of the proton was first revealed by elastic scattering of electrons from protons at electron energies $\sim 1$ GeV, in experiments carried out at Stanford in the early 1960s. The astonishing observation that in deep inelastic scattering of electrons from protons the electrons appeared to be interacting with light, free, point-like objects was made at SLAC in 1968. At that time Feynman was developing a picture of hadrons as clusters of weakly bound, yet not free, grains of energy and momentum, partons. The motivation for this was that in soft-hadron-hadron collisions, the plethora of particles produced are collimated with transverse momenta of typically a few hundred MeV only, relative to the line of flight: in some sense the hadrons fall apart.

The identification of partons with quarks (and gluons) was the culmination of extensive studies of deep inelastic scattering of electrons, muons and neutrinos—lepton probes.

### 6.2 Coulomb scattering and form factors.

Consider elastic scattering of a (spinless) electron from a charged particle, sufficiently massive that it is slow moving in the centre of mass. The scattering then takes place through the coulomb interaction and we can calculate the cross section from a transition matrix element

$$\begin{aligned} \mathcal{M} &= \int e^{-i\mathbf{p}'.\mathbf{r}} \frac{e^2}{r} e^{i\mathbf{p}.\mathbf{r}} d^3r \\ &= \int e^{-i\mathbf{q}.\mathbf{r}} \frac{e^2}{r} d^3r \end{aligned} \tag{6.1}$$

$$\begin{aligned} \mathbf{q} &= \mathbf{p}' - \mathbf{p} \\ q &= 2p \sin\frac{\theta}{2} \end{aligned} \tag{6.2}$$

$$\begin{aligned} \mathcal{M} &= \int e^{-iqr\cos\chi} \frac{e^2}{r} 2\pi r^2 dr\, d\cos\chi = 4\pi e^2 \int_0^\infty \frac{\sin qr}{q} dr \\ &= \frac{4\pi e^2}{q^2} \int_0^\infty \sin x dx \rightarrow \frac{4\pi e^2}{q^2} \end{aligned} \tag{6.3}^*$$

The virtual photon propagator is $\frac{1}{q^2}$ and this virtual photon is not only off mass shell, but $\mathbf{E}.\mathbf{q} \neq 0$:(6.3) is represented by fig.6.1.

---

* The integral $\int_0^\infty \sin dx$ is undefined, but may be controlled by multiplying the potential by a factor $e^{-\epsilon r}$ and taking the limit $\epsilon \rightarrow 0$ after integration. The pure coulomb potential will be shielded at some radius, and electron beams are not perfect plane waves.

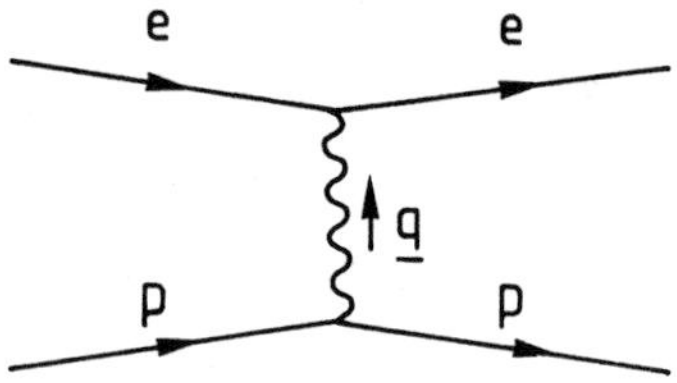


Fig.6.1

Applying the Golden Rule,

$$d\sigma = \frac{2\pi}{v}|\mathcal{M}|^2 d\rho \tag{6.4}$$

$$d\rho = \frac{p^2 dp}{(2\pi)^3 dE} 2\pi d\cos\theta, \; dp/dE = \frac{1}{v}$$

so

$$\frac{d\sigma}{d\Omega} = \left(\frac{4\pi e^2}{q^2}\right)^2 \frac{1}{(2\pi)^2}\frac{p^2}{v^2} \rightarrow \frac{\alpha^2}{4p^2}\operatorname{cosec}^4\frac{\theta}{2} \tag{6.5}$$

as $v \rightarrow 1$. (Natural units $\hbar = c = 1$.) The dimensions are correct, for $d\sigma/d\Omega$ has dimensions GeV$^{-2}$, and this is just the Rutherford scattering formula.

If the charge $e$ were spread out over some volume of space, then (classically) the potential is modified

$$\frac{e^2}{r} \rightarrow e^2 \int \frac{\rho(\mathbf{r}')}{|\mathbf{r}-\mathbf{r}'|} d^3r' \tag{6.6}$$

and

$$\mathcal{M} \rightarrow e^2 \int e^{-i\mathbf{q}.\mathbf{r}} \int \frac{\rho(\mathbf{r}')}{|\mathbf{r}-\mathbf{r}'|} d^3r' d^3r \tag{6.7}$$

Set

$$\mathbf{r} - \mathbf{r}' = \mathbf{R} \quad d^3r \rightarrow d^3R \qquad \text{and then}$$

$$\mathcal{M} \rightarrow \int e^{-i\mathbf{q}.\mathbf{R}} \frac{e^2}{R} d^3R \int e^{-i\mathbf{q}.\mathbf{r}'} \rho(\mathbf{r}') d^3r'$$

or

$$\mathcal{M} = \frac{4\pi e^2}{q^2} F(q^2) \quad ; \quad F(q^2) = \int e^{-i\mathbf{q}.\mathbf{r}'} \rho(\mathbf{r}') d^3r' \tag{6.8}$$

$F(q^2)$ is the form factor—the Fourier transform of the charge distribution. Then

$$\frac{d\sigma}{d\Omega} \rightarrow \frac{\alpha^2}{4p^2}\operatorname{cosec}^4\frac{\theta}{2} F^2(q^2) \tag{6.9}$$

and $F(q^2)$ falls off rapidly with increasing $q^2$. The scale is set by the size of the source. Suppose the charge distribution were gaussian

$$\rho(\mathbf{r}) \simeq e^{-\frac{r^2}{2\sigma^2}} \quad \text{(unnormalised)}$$

Then

$$F(q^2) = e^{-q^2\sigma^2/2}$$

$$F^2(q^2) = e^{-q^2\sigma^2}$$

If $q\sigma \ll 1$ then the scattering is not very sensitive to the size and shape of the source density $\rho(\mathbf{r})$: to resolve structure on a scale $\sigma$ it is necessary that $q\sigma \gtrsim 1$. This is the real criterion for electron microscopy of hadrons. (See Ch.1.)

Suppose $\rho(\mathbf{r}) = \frac{1}{8\pi\sigma^3}e^{-r/\sigma}$ (which is properly normalised)
Then

$$\begin{aligned}
F(q^2) &= \int e^{-i\mathbf{q}.\mathbf{r}}\frac{1}{8\pi\sigma^3}e^{-r/\sigma}d^3r \\
&= \int e^{-iqr\cos\chi}\frac{1}{8\pi\sigma^3}e^{-r/\sigma}2\pi r^2 dr\, d\cos\chi \\
&= \int_0^\infty \frac{[e^{iqr}-e^{-iqr}]}{iq}e^{-r/\sigma}\frac{rdr}{4\sigma^3} \\
&= \frac{1}{(1+q^2\sigma^2)^2} \qquad (6.10)
\end{aligned}$$

This is in fact what is observed. Things are not this simple—we have neglected the complications of spin and in particular magnetic moment scattering. But both coulomb and magnetic moment scattering of electrons from the proton behave as though the scattering comes from little bits of spin 1/2 distributed exponentially inside the proton. Magnetic moment scattering from the neutron behaves in the same way. For the nucleon

$$< r^2 >= (0.8\text{fm})^2$$

Since $< r^2 >= 12\sigma^2 \qquad \sigma = 0.23\text{fm}$.

If the nucleon form factor were due to intermediate vector mesons (fig.6.2)

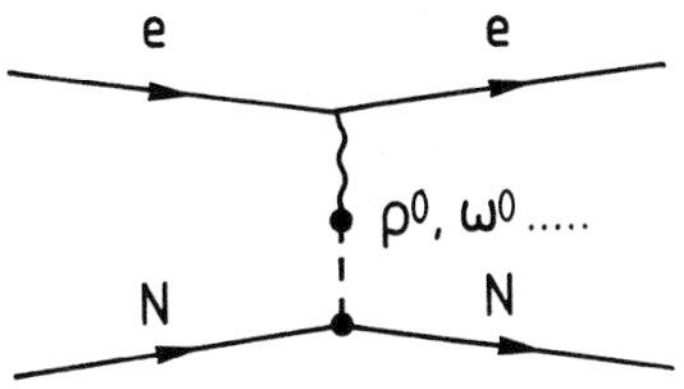


Fig.6.2

we would expect a form factor from the meson propagator $\approx \frac{1}{m_v^2+q^2}$ and not the so called dipole form which is actually observed, (6.10).

Similar considerations apply to neutrino 'elastic' scattering

$$\begin{aligned}
\nu_\mu + n &\rightarrow \mu^- + p \\
\bar{\nu}_\mu + p &\rightarrow \mu^+ + n
\end{aligned}$$

Spin matters more here: the weak interactions violate parity. Ignoring this complication, if we have a genuine point interaction

$$\sigma \sim 2\pi G^2\rho = 2\pi G^2\frac{4\pi p^2}{(2\pi)^3} = \frac{G^2p^2}{\pi} \qquad (6.11)$$

and this cross section grows as the centre of mass momentum squared—linearly with the laboratory neutrino energy. If the neutrinos interact with little bits distributed within the proton then we have for the matrix element

$$G \to G \int e^{-i\mathbf{q}.\mathbf{r}} \rho(\mathbf{r}) d^3 r = GF(q^2) \tag{6.12}$$

$F(q^2)$ is experimentally the same for neutrino scattering as for electron scattering. Then

$$d\sigma \sim 2\pi G^2 F^2(q^2) \frac{p^2}{(2\pi)^3} 2\pi d\cos\theta \tag{6.13}$$

Remember that $p^2 d\cos\theta = \frac{1}{2} dq^2$ (see eqs.2.11). Then

$$\sigma = \frac{G^2}{4\pi} \int \frac{dq^2}{(1+\sigma^2 q^2)^4} \sim \frac{G^2}{12\pi\sigma^2} \tag{6.14}$$

The total elastic cross section flattens off as soon as $\sigma q_{\max} \gg 1$ —above $\sim 1$ GeV in the centre of mass.

$$\sigma \sim \frac{G^2}{12\pi\sigma^2} = \frac{[10^{-5}\,\mathrm{GeV}^{-2}]^2}{12\pi[0.23\,\mathrm{fm} \times 5\,\mathrm{GeV}^{-1}]^2} \approx \frac{10^{-11}}{3}\,\mathrm{GeV}^{-2}$$

In more familiar units $\sigma \sim 10^{-13}\,\mathrm{fm}^2 = 10^{-12}\,\mathrm{mb} = 10^{-39}\,\mathrm{cm}^2$.

### 5.3 Deep inelastic scattering.

The surprises came in when deep inelastic lepton scattering was studied. This means the processes

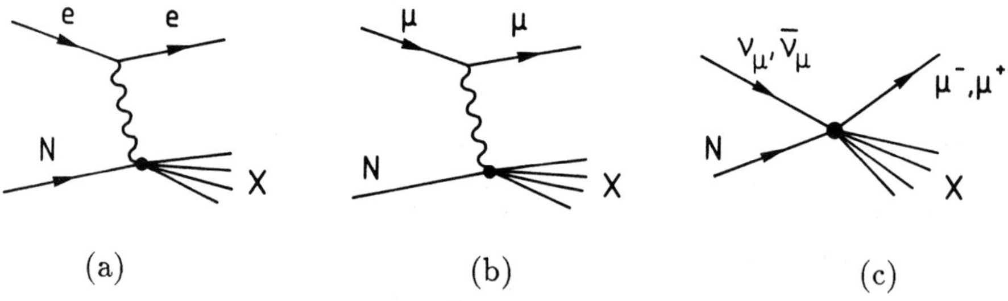


Fig.6.3

where in the centre of mass the recoiling lepton does not have the same energy as the incident lepton. The virtual photon (or $W$) carries substantial energy as well as momentum (but $q^2$ is still positive—a frame exists in which only momentum is transferred). These are called inclusive processes—because $X$ is anything nature likes.

Above a laboratory energy of a few GeV, inclusive neutrino cross sections continue to rise linearly with laboratory energy up to at least 100 GeV. There is no $q^2$ dependent form factor in the inclusive process: it is just as though neutrinos interact at a point with little point like bits inside the nucleon, which get knocked out. [Note that even at 100 GeV, $E_{\mathrm{cm}} \sim 20$ GeV $\ll M_W$]

In deep inelastic scattering of electrons (or muons) the only $q^2$ dependence is the familiar (Rutherford) $1/q^4$ behaviour due to the virtual photon propagator—and this behaviour sets in at 4-momentum transfers of only a few GeV. Again there is no $q^2$ dependent form factor. Vector meson dominance would imply a similar form factor for inelastic processes as for elastic.

How might we have tried to explain what goes on when CONFINED little bits are struck by a lepton? For coulomb scattering we started with a matrix element

$$e^2 \iint e^{-i\mathbf{q}.\mathbf{r}} \frac{\rho(\mathbf{r}')}{|\mathbf{r}-\mathbf{r}'|} d^3r d^3r' \tag{6.7}$$

where $\rho$ was a classical charge density. For elastic scattering we replace $\rho$ by a quantum mechanical probability density

$$\rho(\mathbf{r}') \rightarrow \Psi^*(\mathbf{r}')\Psi(\mathbf{r}') \tag{6.15}$$

where $\Psi(\mathbf{r}')$ is the wave function for the little bits carrying the charge. Since we know excited states exist, the obvious generalisation for inelastic scattering is

$$\rho(\mathbf{r}') \rightarrow \Psi_x^*(\mathbf{r}')\Psi(\mathbf{r}') \tag{6.16}$$

For confined little bits both $\Psi_x$ and $\Psi$ are localised so a form factor

$$\int e^{-i\mathbf{q}.\mathbf{r}'} \Psi_x^*(\mathbf{r}')\Psi(\mathbf{r}')d^3r'$$

is expected to be a strong function of $q^2$ once more. Thus as energy transfer increases one expects spikes in the cross section as a function of energy transfer at fixed angle, corresponding to recoiling excited states, but the cross section for each excited state should fall off rapidly as a function of $q^2$. This is observed for small energy transfer; excitation of the first few states of the nucleon is quasi-elastic scattering rather than deep inelastic scattering.

The reason for the strong $q^2$ dependence expected is the localised nature of the excited states in momentum space. The probability of a little bit which has been given a hefty kick overlapping the tail of the momentum distribution of a little bit in an excited state should fall off very rapidly with the momentum transfer. In an atom or a nucleus, of course, as soon as the energy transfer exceeds the ionisation potential (or nucleon separation energy) a struck electron or nucleon behaves as though it is free, because it is transferred to the continuum where a suitable set of eigenstates are plane waves with all momenta. The little bits within the nucleon (quarks) are not observed propagating freely. The contradiction is resolved provided that, for energy transfers exceeding $\sim$ GeV, any momentum state of the struck quark generates a physically accessible hadronic final state. In the string picture this happens because with $\sim$ 1 GeV/fm stored in the string, a string with a potential length of several fermi readily breaks up into hadrons—which do propagate freely—before the excited string has completed even half an oscillation.

The recipe in general is:

(i) Transfer energy-momentum to a quark

(ii) Measure the overlap of the wave function resulting with the spectrum of allowed physical final states.

The first step is calculated with electromagnetic or weak interactions of point like particles. The second has an effect depending on the kinematics of the first. Pictorially:

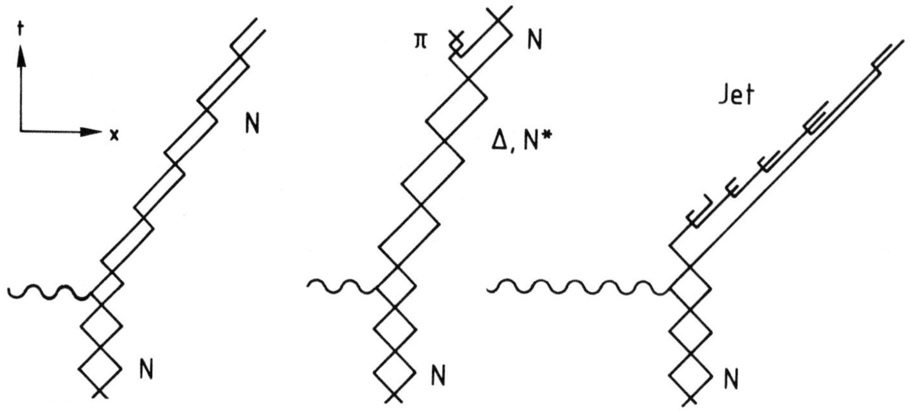


Elastic scattering (a) Quasi-elastic scattering (b) Deep inelastic scattering (c)

Fig.6.4

We have here combined a classical string picture with quantum mechanical arguments. Classically string states form a continuum and all masses are possible. If the mass of the hadronic system exceeds $\sim 2$ GeV, the picture shown in fig.6.4(c) is acceptable and useful. The final state shown in fig.6.4(a) has a unique mass, and that of fig.6.4(b) is relatively narrow. In the semi-classical string model it is possible to accept a band of classical masses as representing the nucleon, another as representing a particular excited state. While it is possible to quantise the two dimensional string using a Feynman sum over histories, I do not know how to calculate form factors in such a model.

Thus in deep inelastic scattering the cross section is that of leptons scattering from quarks, contained in the nucleon, which behave like free particles. You may worry that the size of the nucleon sets some kind of scale in the cross section because of the initial quark wave function. This scale controls the distribution of momentum of the bits that were not struck and is integrated away when summing over all possible final states, (Problem 6.3). The only effect of the quarks being confined within a nucleon initially is that while (in the target frame) the sum of quark energies is fixed, they have a spread of momenta, controlled by the space structure of the wave function (through a Fourier transform). In a frame in which the nucleon has infinite momentum, this distribution determines the distribution of fractional nucleon momentum carried by the quarks. It is this distribution which is called the quark structure function of the nucleon, and is much more fundamental than (non-relativistic) wave functions. We should one day be able to calculate structure functions from QCD, in principle.

### 6.4 The crossed channels.

It is obvious that the above representation of deep inelastic scattering has much in common with $e^+e^-$ annihilation into hadrons. Under crossing fig.6.3(a) becomes

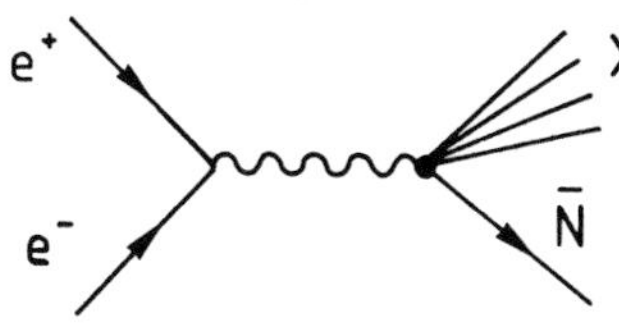


Fig.6.5

which is a particular case of hadron production in $e^+e^-$ annihilation. [The debris $X$ must contain a baryon.] The resemblance extends to stable and long lived quark states, for at the appropriate energies individual bound states dominate $e^+e^-$ annihilation. Two examples are shown in fig.6.6, the second of which is famous

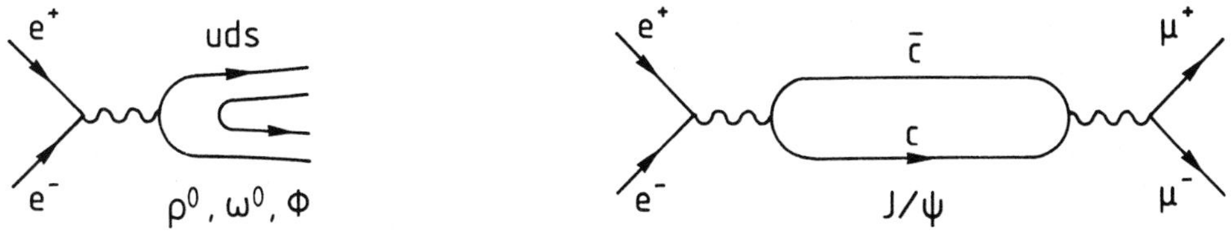


Fig.6.6

In the string picture these processes are represented by fig.6.7

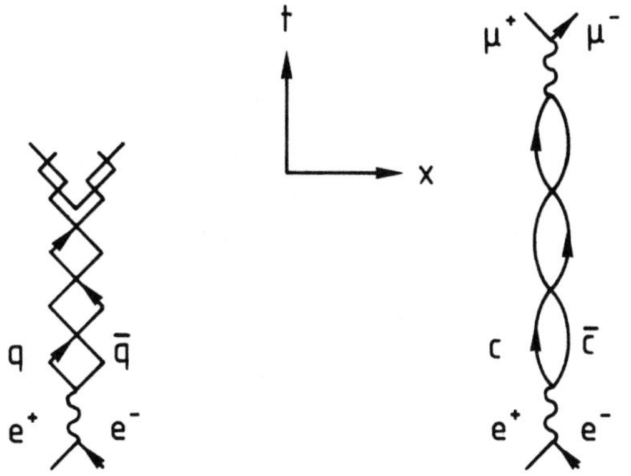


Fig.6.7

You see experimentally big increases in the cross sections—resonances—when the energy is just right to make the vector mesons $\rho^\circ$, $\omega^\circ$, $\phi$ near 1 GeV and $J/\psi$ at 3.1 GeV (Problem 6.4). $J/\psi$ is very narrow. It is a $c\bar{c}$ bound state below threshold for decay into two mesons each containing a charmed quark (just as $\phi$ is almost below threshold for $K\bar{K}$—but not quite). It decays into hadrons through the disconnected diagram

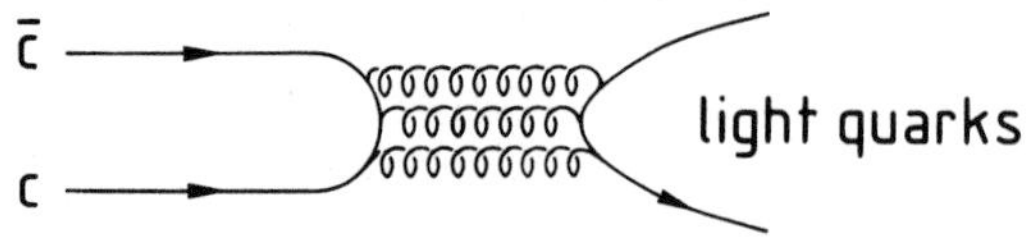


Fig.6.8

The minimal colour field exchange has the quantum numbers of three gluons.

A single gluon is coloured, and hadronic states are colourless. The colour exchange must be at least two gluons. The $J/\psi$ is a triplet $s$-wave state of $c\bar{c}$, and so under the charge conjugation operation ($C$) $J/\psi$ is negative. [Remember that a fermion-antifermion state has negative intrinsic parity.] The colour field is a vector field coupling to a conserved current, like electromagnetism, and so changes sign under the $C$ operation. Finally, $C$ is multiplicative and so a gluon state of odd $C$ must be built from an odd number of gluons. It is presumed that $J/\psi$ is narrow (the width is 0.06 MeV yet the dominant decay is to hadrons) because of the high order of fig.6.8.

### 6.5 The Drell-Yan mechanism.

There is another entertaining lepton probe of quarks within hadrons—the Drell-Yan mechanism. In a sense it is the converse of $e^+e^-$ annihilation:

$$q\bar{q} \to e^+e^-, \quad q\bar{q} \to \mu^+\mu^-$$

in hadron-hadron collisions. The complete cross section for the process

$$hh \to \mu^+\mu^- + X$$

can in principle be calculated provided the hadron structure functions are known (and conversely this is the only way to get at the structure functions of mesons)

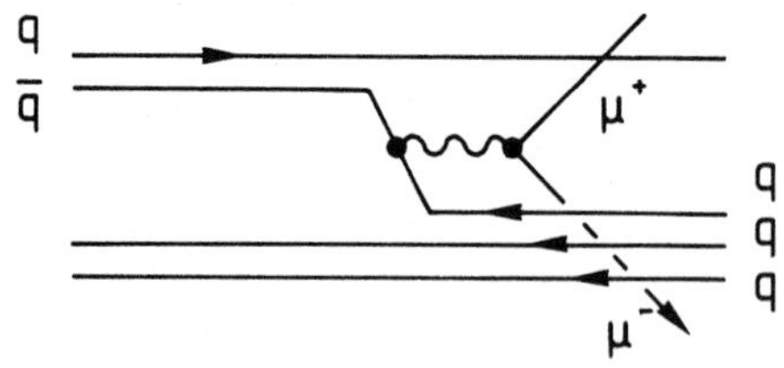


Fig.6.9

Obviously there will be hadronic debris—the quark annihilation leaves a triplet-anti triplet of colour flying apart and a string will stretch between them. Relative cross sections are rather easy to calculate:

| | | | | | | | |
|---|---|---|---|---|---|---|---|
| $\pi^+p$ | $u\bar{d}$ | $duu$ | $1 \times \left(\frac{1}{3}\right)^2$ | $\pi^+n$ | $u\bar{d}$ | $ddu$ | $2 \times \left(\frac{1}{3}\right)^2$ |
| $\pi^-p$ | $d\bar{u}$ | $uud$ | $2 \times \left(\frac{2}{3}\right)^2$ | $\pi^-n$ | $d\bar{u}$ | $udd$ | $1 \times \left(\frac{2}{3}\right)^2$ |

On an 'isoscalar' target (C, Fe, Cu ...)

$$\frac{\pi^+ N}{\pi^- N} = \frac{1/9 + 2/9}{8/9 + 4/9} = \frac{1}{4}$$

and it is, in the appropriate kinematic region. For $\bar{p}p$ the sum over pairs of quarks gives $1\times\left(\frac{1}{3}\right)^2 + 4\times\left(\frac{2}{3}\right)^2 = 17/9$. In $pp$ the same argument yields zero. However, we do observe $\mu^+\mu^-$ pairs in $pN$ interactions, at a rate $\sim 10^{-2}$ of $\bar{p}N$ interactions. The hadrons contain in addition to valence quarks $q\bar{q}$ pairs produced transiently through quantum fluctuations. This is a piece of information also extracted from neutrino interactions (Ch.10).

Incidentally, when $q\bar{q}$ energy in the $(q\bar{q})$ centre of mass is right, processes such as those of fig.6.10 are observed

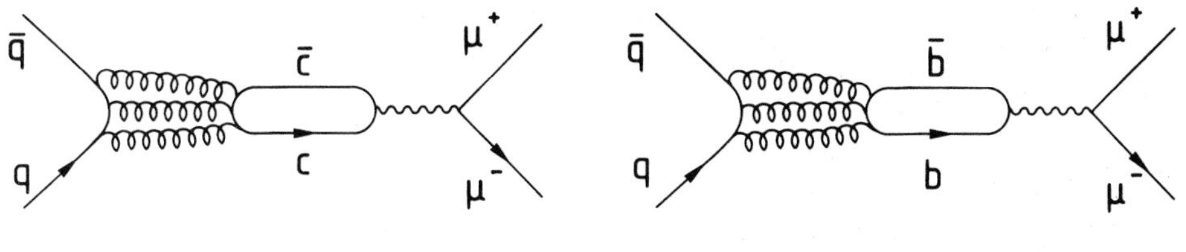


Fig.6.10

The $J/\psi$ was discovered in $pN$ collisions in this way, simultaneously with its discovery in $e^+e^-$ annihilation. The first observations of the $\Upsilon(b\bar{b})$ states were made in $pN \rightarrow \mu^+\mu^-$.

Hadron-hadron colliders act as quark-quark, quark-gluon and gluon-gluon colliders, as far as hard processes are concerned; those in which particles or jets with momentum transverse to the beams in excess of a few GeV are produced. The present generation of hadron colliders contains only the CERN $p\bar{p}$ collider (540 GeV centre of mass energy, although the machine has been cycled up and down in energy to reach a maximum of 900 GeV) and the Fermilab superconducting Tevatron collider, which has reached energies in excess of 1600 GeV in the centre of mass. The $W^\pm$ were first produced at CERN in $u\bar{d}$, $\bar{u}d$ collisions. Quark-quark, quark-gluon and gluon-gluon elastic scattering should proceed predominantly through the exchange of a single virtual gluon and the resulting high transverse momentum jets should have, in the jet-jet centre of mass, the angular distribution characteristic of Rutherford scattering. They do.

The representation of the long range colour interactions in terms of a tube of chromoelectric flux, a colour string, makes it possible to understand how quarks can behave as if free in deep inelastic scattering and $e^+e^-$ annihilation: the phenomenon called asymptotic freedom. [There is one proviso: the rate for the radiation of hard gluons must not be too large.] It also provides a simple physical model for colour confinement, or infrared slavery. If a hard gluon is radiated in either deep inelastic scattering or $e^+e^-$ annihilation, the gluon can be represented as a kink on the string, carrying energy and momentum. The colour string picture is not a theory, and as a phenomenological model it is far from complete. It has however been remarkably suggestive and remarkably successful.

## Problems

6.1 Assuming a spherically symmetric distribution of charge, show that the form factor is a function of $q^2$ and that for sufficiently small values of $q^2$

$$F(q^2) \simeq 1 - \frac{1}{6} q^2 < r^2 >$$

where $< r^2 >$ is the mean square radius of the distribution.

Consider the elastic scattering of 1 GeV electrons by protons (at rest in the laboratory). Calculate the maximum value of $q^2$ attainable. Calculate the values of $q^2$ corresponding to laboratory scattering angles of 2°, 30° and 90°. Compare the ratios of the number of electrons coulomb scattered into equal intervals of $q^2$ centred on each of these three values, assuming a point proton. Calculate the modification to these ratios due to the dipole proton charge form factor.

6.2 An electron (or any other particle) is elastically scattered from a target of mass $M$ which is initially at rest in the laboratory. The target recoils with four-momentum $(\mathbf{p}, \nu)$. Show that the square of the four-momentum transfer satisfies the relation

$$\frac{q^2}{2M\nu} = 1$$

Obviously $(\mathbf{p}, \nu)$ can be constructed from the initial and final electron momenta in the laboratory, for $-\mathbf{p}$ is the momentum transferred and $M - \nu$ is the energy transferred from the incident electron to the scattered electron; for relativistic recoil the energy transferred to the scattered electron is $-\nu$.

In deep inelastic scattering from nucleons of mass $M$ the cross section may be determined as a function of the momentum and energy transferred to the scattered electron, and $\nu$ is defined to be the energy lost by the electron, in the laboratory frame. The cross section is that for point-like scattering at momentum transfer squared $q^2$, modified by a factor $f\left(\frac{q^2}{2M\nu}\right)$. This function peaks at $\frac{q^2}{2M\nu} \sim \frac{1}{3}$.

Draw the obvious inferences. (They were not obvious in 1968). [The quantity $x = \frac{q^2}{2M\nu}$ is known as the Bjorken scaling variable.]

6.3* This is a simple but instructive exercise in non-relativistic quantum mechanics. Consider the scattering of an electron from a deuteron, where the energy is sufficiently high that we are interested only in the breakup of the deuteron via coulomb scattering of the electron from the bound proton. The energies of the nucleons are to be neglected. Suppose that the final state consists of an electron with momentum $\mathbf{p}_1'$, a proton with momentum $\mathbf{p}_2'$ and a neutron with momentum $\mathbf{p}_3'$—plane waves. The initial state consists of an electron with momentum $\mathbf{p}_1$ and a deuteron at rest. The electron coordinate is $\mathbf{r}_1$, the proton and neutron are at $\mathbf{r}_2$ and $\mathbf{r}_3$. The wave function describing proton-neutron separation in the initial bound state is taken to be

$$\psi_D = \sqrt{\frac{\beta^3}{\pi}} e^{-\beta r} \qquad r = |\mathbf{r}_2 - \mathbf{r}_3|$$

Construct the matrix element and show that the deuteron structure induces no $q^2$ dependence in this process. Show further that on integration over the final momenta of the spectator neutron, the parameter $\beta$ vanishes from the cross section.

6.4 The lowest lying $c\bar{c}$ state is the $J/\psi$ at a mass of 3.097 GeV, which is below threshold for charm production. [The curious and unwieldy name commemorates simultaneous discovery in proton-nucleus interactions (Brookhaven) and $e^+e^-$ annihilation (SLAC).] The total width (full width at half height) is 0.063 MeV and the partial widths to $e^+e^-$, $\mu^+\mu^-$ are both 4.4 KeV. Calculate the resonant cross sections at the $J/\psi$ for $e^+e^- \to \mu^+\mu^-$ and $e^+e^- \to$ hadrons. Compare the peak cross sections with the non-resonant cross sections at the same energy. [You may then imagine the reactions of the physicists who had the watch when the SPEAR energy scan reached almost 3.1 GeV for the first time].

The width of the $J/\psi$ is small in comparison with the energy spread in the beams of an $e^+e^-$ collider. Show that the total width and the branching ratios could nonetheless be determined from the integrals of cross section with respect to energy.

6.5 The structure functions of nucleons give the probability that a particular class of parton carries a given fraction $x$ of the nucleon momentum, strictly speaking in a frame where the nucleon momentum is infinite. They have been determined in deep inelastic scattering of electrons, muons and neutrinos. Convince yourself that the valence quark structure functions extracted are independent of whether or not colour is attached to the quarks, bearing in mind that hadrons are colour singlet states. The Drell-Yan cross sections can in principle be determined, given the structure functions. Show that if quarks are coloured the Drell-Yan cross sections have one third of the value that would obtain without colour. [It is a good idea to consider first deep inelastic scattering from a meson and meson-meson Drell-Yan processes.]

# 7. GOLDEN RULES.

## 7.1 The Fermi Golden Rule.

We have assumed so far that you are acquainted with either the first Born approximation in scattering theory or the derivation of the equivalent result in time-dependent perturbation theory. In this Chapter we treat the Fermi Golden Rule for the first time. The rule is derived in perturbation theory and its validity even when perturbation theory is not applicable is demonstrated. The usual treatment has some apparently dubious aspects, and in 7.2 we treat decays explicitly and unravel the connection between the width of a state, its lifetime and the Breit-Wigner denominator. The invariant matrix elements and invariant phase space generally employed in particle physics are dealt with, although we shall not really need them in subsequent work. We begin with the conventional derivation of the Fermi Golden Rule in first order time-dependent perturbation theory.

In particle physics we are concerned with generalised decay problems and generalised scattering problems (which may be two in, many out). One would like to be able to calculate (at least in principle) decay rates and scattering cross sections. The machinery is encapsulated in the Fermi Golden Rule

$$T_{if} = 2\pi|\mathcal{M}_{if}|^2\rho_f \quad ; \quad \rho_f = dN_f/dE_f \tag{7.1}$$

where $T_{if}$ is the transition rate from $i$ to $f$, $\mathcal{M}$ is a matrix element and $\rho_f$ is the number of states $f$ in unit energy interval. Our final states are part of a continuum however, so what does $\rho$ mean? We calculate with discrete states, of definite energy, and make these by confining our outgoing particles to a box. The number of states between $\mathbf{p}$ and $\mathbf{p} + d\mathbf{p}$ is

$$\frac{p^2\,dp}{(2\pi)^3}d\Omega V \tag{7.2}$$

[units $\hbar = 1$] where $V$ is the volume of the box. The matrix element is taken between states normalised to the same volume, and $V$ cancels out. In counting the number of discrete states we have already assumed $V$ much greater than any wavelength, so we let $V \to \infty$ to reach the continuum and when $\rho$ is evaluated with $V$ set equal to unity it corresponds to the number of states in unit volume, the plane waves being normalised to unit probability density.

If we can use perturbation theory, $\mathcal{M}_{if}$ can be calculated. If not, we usually know something about $\mathcal{M}_{if}$ and the Golden Rule imposes at least the necessary structure.

The usual derivation is as follows. The time evolution of a system is described by a Schrödinger equation

$$H\psi = i\frac{\partial\psi}{\partial t} \tag{7.3}$$

and we divide $H$ into a dominant term $H_0$ and a perturbation $H'$

$$H = H_0 + H' \tag{7.4}$$

The eigenstates of $H_0$ satisfy

$$H_0\psi_n = E_n\psi_n \tag{7.5}$$

and

$$\psi_n = \phi_n e^{-iE_nt} \tag{7.6}$$

where $\phi_n$ is independent of time.

A general $\psi$ is expanded in terms of the complete set $\psi_n$

$$\psi = \sum a_n\psi_n \tag{7.7}$$

when

$$H\psi = i\frac{\partial\psi}{\partial t} \quad \rightarrow \quad \sum_n a_n H'\psi_n = i\sum_m \psi_m \frac{\partial a_m}{\partial t} \tag{7.8}$$

whence

$$\frac{\partial a_m}{\partial t} = \frac{1}{i}\sum a_n \int \psi_m^* H'\psi_n d^3x \tag{7.9}$$

The usual argument then goes as follows. Start the system off with $a_i = 1$, all other coefficients zero. Then if $H'$ is very weak keep all terms fed from initially empty states zero. Also suppose $a_i$ stays $\sim 1$. This is plausible for a scattering problem where we are feeding in a wave from outside—but it is dubious for decay processes, which we treat in 7.2. Then

$$a_f = \frac{1}{i}\int \psi_f^* H'\psi_i d^3x dt \tag{7.10}$$

This is interesting. If $H'$ has no explicit time dependence and the time integral extends from $-\infty$ to $\infty$, then $E_f - E_i \rightarrow 0$—conservation of energy. If $H'$ oscillates with time as $e^{-i\omega t}$ then $E_f = E_i + \omega$—energy is gained from the field driving $H'$. If $\psi_f$ is an electron and we give $\psi_i$ negative energy, an initial negative energy electron is interpreted as a final positive energy positron and $\omega = E_{e^+} + E_{e^-}$. Pair creation is built in as we go relativistic.

We often pretend the limits are $-\infty$ to $\infty$, but in reality we start the process some time. Either switch on $H'$ at $t = 0$ and switch it off at $t$, or perhaps switch on $\psi_i$ at $t = 0$ and switch it off at $t$. This is what really happens in scattering. Then the time part of $a_f$ is

$$\int_0^t e^{iE_ft}e^{-i\omega t}e^{-iE_it}dt$$

[$\omega$ could be the energy of an appropriate Fourier component of $H'$]. We obtain $a_f$ as a function of the energy difference $\Delta E$ by integrating between $t = 0$ and $t$

$$\int_0^t e^{i\Delta Et}dt = (e^{i\Delta Et} - 1)/i\Delta E \tag{7.11}$$

and

$$|a_f|^2 = |<\phi_f^*|H'|\phi_i>|^2 \frac{\sin^2\left(\frac{\Delta Et}{2}\right)}{\left(\frac{\Delta E}{2}\right)^2} \tag{7.12}$$

The probability $|a_f|^2$ that a state $f$ is occupied after a time $t$ is very sharp in terms of the energy difference $\Delta E$ if $t$ is large.

For a very big box, we are interested in the transition rate into states $f$ such that $\Delta E \sim 0$ so we want the transition rate

$$\frac{d}{dt}\sum_f |a_f|^2 \to \frac{d}{dt}\int |a_f|^2 \rho(E_f) dE_f \tag{7.13}$$

In most cases $t$ will be large, so we can be sure the band of energies is very narrow and take both $|<\phi_f^* H'\phi_i>|^2$ and $\rho(E_f)$ out of the integral. With the replacement $dE_f \to d\Delta E$ we may extend (for a narrow band) the limits on $\Delta E$ from $-\infty$ to $\infty$ and then we have

$$\begin{aligned}\int_{-\infty}^{\infty} \frac{d}{dt}\frac{\sin^2 \frac{\Delta E}{2}t}{\left(\frac{\Delta E}{2}\right)^2} d\Delta E &= \int_{-\infty}^{\infty} \frac{4\sin\frac{\Delta Et}{2}\cos\frac{\Delta Et}{2}\frac{d\Delta E}{2}}{\frac{\Delta E}{2}} \\ = \int_{-\infty}^{\infty} 4\frac{\sin x \cos x}{x} dx &= 2\pi\end{aligned} \tag{7.14}$$

and hence

$$T_{i\to f} = 2\pi \left|<\phi_f^*|H'|\phi_i>\right|^2 \rho(E_f) \tag{7.15}$$

where $\rho(E_f)$ is the density of specified final states in terms of $E_f$. In a generalised scattering problem the cross section is obtained by dividing the rate $T_{i\to f}$ by a flux factor. This is just the relative velocity of the (two) particles making up $\phi_i$—a normalisation volume $V$ cancels again.

The transition rate can be written in this form regardless of whether the matrix element can be calculated in perturbation theory. In an (approximately) steady state scattering problem incident waves go in and scattered waves emerge. They do not interact once they have left the region where the interaction rate is significant, and they never come back. If we are referring to such asymptotic states in a scattering process the argument is generally valid given some operator $\mathcal{M}$ which acting on an asymptotic incoming state yields (by all sorts of complicated processes if you like) an outgoing asymptotic state $\psi_f$. Then we really can write

$$\frac{\partial a_f}{\partial t} = \frac{1}{i}\int \psi_f^* \mathcal{M}\psi_i d^3x$$

and the Golden Rule follows.

## 7.2 Decaying states.

The problem of a decaying state has some interesting aspects. Again we most interested in asymptotic outgoing states which do not talk to each other.

Suppose we have two states—an initial state 1 and a particular final state 2. Then

$$\psi = a_1\psi_1 + a_2\psi_2 = A_1(t)\phi_1 + A_2(t)\phi_2$$

where all time dependence is contained in the functions $A(t)$. Rewrite the Schrödinger equation in the form

$$\begin{aligned} i\frac{\partial A_1}{\partial t} &= A_1 < \phi_1^*|H|\phi_1 > + A_2 < \phi_1^*|H|\phi_2 > \\ i\frac{\partial A_2}{\partial t} &= A_1 < \phi_2^*|H|\phi_1 > + A_2 < \phi_2^*|H|\phi_2 > \end{aligned} \tag{7.16}$$

Solve—just as for coupled oscillators—with $e^{\lambda t}$

$$i\lambda A_1 = A_1 H_{11} + A_2 H_{12} \tag{7.17(a)}$$

$$i\lambda A_2 = A_1 H_{21} + A_2 H_{22} \tag{7.17(b)}$$

This gives two solutions for $\lambda$: $\lambda_1 \sim -iH_{11}$ and $\lambda_2 \sim -iH_{22}$ in perturbation theory. For asymptotic states we want the limit in which $H_{12}$ (and in general $H_{1m}$) is inoperative and of course the states $2 \ldots m$ do not affect one another. Now from 7.17(b)

$$A_2 = \frac{A_1 H_{21}}{i\lambda - H_{22}}$$

so that we can write

$$A_1(t) = A_1^1 e^{\lambda_1 t} + A_1^2 e^{\lambda_2 t} \tag{7.18(a)}$$

$$A_2(t) = \left\{ \frac{A_1^1 e^{\lambda_1 t}}{i\lambda_1 - H_{22}} + \frac{A_1^2 e^{\lambda_2 t}}{i\lambda_2 - H_{22}} \right\} H_{21} \tag{7.18(b)}$$

As we switch off $H_{12}$ $\quad i\lambda_2 \to H_{22}$ $\quad$ and $\quad A_1^2 \to 0$ $\quad$ such that $A_2(0) = 0$ and $A_1(0) = 1$. Then

$$A_1(t) \to e^{\lambda_1 t} \tag{7.19(a)}$$

$$A_2(t) \to \frac{H_{21}}{i\lambda_1 - H_{22}} \left\{ e^{\lambda_1 t} - e^{\lambda_2 t} \right\} \tag{7.19(b)}$$

Since nothing comes back and with a decaying state we are not feeding the system from outside, $\lambda_1$ contains a negative real part. Then as $t \to \infty$

$$\begin{aligned} |A_2(t)|^2 &\to \frac{|H_{21}|^2}{|i\lambda_1 - H_{22}|^2} \\ |A_f(t)|^2 &\to \frac{|H_{f1}|^2}{|E_1 - i\frac{\Gamma}{2} - E_f|^2} \end{aligned} \tag{7.20}$$

for all final states $f$. For conservation of probability (also known as unitarity) the sum of final state probabilities must be equal to unity

$$\sum |A_f(\infty)|^2 \to 1$$

or

$$\int \frac{|H_{f_1}|^2}{(E_1 - E_f)^2 + \frac{\Gamma^2}{4}} \rho(E_f) dE_f = 1 \qquad (7.21)$$

Take $|H_{f_1}|^2 \rho(E_f)$ outside the integral and we have

$$\rho(E_f)|H_{f_1}|^2 \int \frac{dE_f}{(E_1 - E_f)^2 + \frac{\Gamma^2}{4}} = 1$$

$$\frac{2\pi}{\Gamma}\rho(E_f)|H_{f_1}|^2 = 1 \qquad \Gamma = 2\pi\rho(E_f)|H_{fi}|^2 \qquad (7.22)$$

and of course $|A_1(t)|^2 = e^{2\,\mathrm{Re}\,\lambda_1 t} = e^{-\Gamma t}$. The decay rate is thus $\Gamma = \frac{1}{\tau}$ and $\Gamma\tau = 1$. We have again obtained the Fermi Golden Rule.

Note that as we ignore $H_{12} \quad i\lambda_1 \to H_{11}$ , so $H_{11}$ has an imaginary part. This is important if we consider scattering through a state instead of preparing it at $t = 0$ and then watching it decay.

### 7.3 Intermediate states.

We can extend the treatment to a higher order process in which the first step is formation of an intermediate state, or resonance, and this state subsequently decays to two or more particles which emerge as asymptotic plane waves.

Suppose we get from 1 to 3 only by passing through the unstable state 2. Maintain 1 at 1/unit volume, with energy $E$. Then

$$i\frac{\partial A_2}{\partial t} = H_{21}A_1 + H_{22}A_2 \qquad (7.23)$$

where $H_{22}$ contains an imaginary piece because the state is decaying. For steady state conditions, after all transients have died out,

$$A_2 = \frac{H_{21}A_1}{i\lambda - H_{22}} \to \frac{H_{21}}{E_1 - E_2 + i\frac{\Gamma_2}{2}} \qquad (7.24)$$

$$|A_2|^2 = \frac{|H_{21}|^2}{(E_1 - E_2)^2 + \frac{\Gamma_2^2}{4}} \qquad (7.25)$$

and state 2 decays at a rate $\Gamma_{2f}|A_2|^2$ to states $f$ where $\Gamma_{2f} = 2\pi|H_{f2}|^2\rho(E)$. The transition rate to all states $f$, of which 3 was a representative member, is

$$T_{1\to2\to f} = \frac{2\pi|H_{f2}H_{21}|^2}{(E - E_2)^2 + \frac{\Gamma^2}{4}}\rho(E) \qquad (7.26)$$

where $\Gamma$ is the total width, the sum of the partial decay rates to all possible channels. The cross section is obtained by dividing by a flux factor, and we obtain the standard Breit-Wigner resonance formula giving the variation of the cross section with centre of mass energy when the energy spread of the beam is $\ll \Gamma$. In all these discussions involving asymptotic states, $H_{ij}$ are not in general matrix elements of the Hamiltonian, but are approximated by matrix elements of the Hamiltonian when the interactions are weak and perturbation theory applies. $H_{21}$ may be replaced by an expression involving $\Gamma_{12}$ and $\Gamma_{ij}$ may be treated as a set of parameters or functions characteristic of a resonance. They are not in general energy independent.

### 7.4 Propagators and resonances.

In Ch.2 we considered the scattering of two particles through a static Yukawa potential of the form

$$V(r) = g^2 \frac{e^{-mr}}{r} \tag{7.27}$$

where $m$ is the mass of the free field. The matrix element took the form

$$\mathcal{M}(\mathbf{p}_i \to \mathbf{p}_f) = \frac{4\pi g^2}{q^2 + m^2} \tag{7.28}$$

where $q^2 = (\mathbf{p}_f - \mathbf{p}_i)^2$ is the square of the momentum transferred by the field, in the centre of mass. A virtual boson carrying momentum $\Delta\mathbf{p} = \mathbf{p}_f - \mathbf{p}_i$ was created at one vertex and absorbed at the other. For elastic scattering, no energy is transferred in the centre of mass. The momentum transfer squared, defined in the centre of mass, is a Lorentz invariant

$$q^2 = \Delta p^2 - \Delta E^2$$

and we can create and destroy virtual particles carrying arbitrary momentum and energy. If two particles annihilate, then in the centre of mass

$$q^2 = -\Delta E^2 = -E^2$$

and if (7.28) is taken as the matrix element for this process, it looks very like the resonance amplitude (7.24)

$$\mathcal{M}(E) = \frac{-4\pi g^2}{E^2 - m^2} \tag{7.29}$$

except that the denominator contains

$$E^2 - m^2 = (E - m)(E + m)$$

instead of the term $E - m$ appropriate to (7.24).

Remember that the covariant field equation (for a scalar field) is

$$\left(\nabla^2 - \frac{\partial^2}{\partial t^2}\right)\phi - m^2\phi = S(x,t) \tag{7.30}$$

where $S$ is a source term which is also scalar. The field $\phi$ may be Fourier transformed into momentum space to yield

$$\tilde{\phi} = -\frac{\tilde{S}}{q^2 + m^2} \tag{7.31}$$

where $\tilde{S}$ is the Fourier transform of $S$ and is a constant for a point source in $(x, t)$. If $q^2$ is positive momentum is transferred, whereas if $q^2$ is negative the field transfers energy with a probability amplitude $\sim 1/(E^2 - m^2)$. For a free field we have plane waves subject to the condition $q^2 + m^2 = 0$.

The calculation which led to (7.24–7.26) employed a (non-relativistic) Hamiltonian with strict conservation of momentum and strict time order. Energy was not conserved—the intermediate state exists on a short timescale. The non-relativistic treatment took no account of antiparticles and the creation and annihilation of pairs, and in order to take them into account we should add to the amplitude of fig.7.1(a) the amplitude of fig.7.1(b)

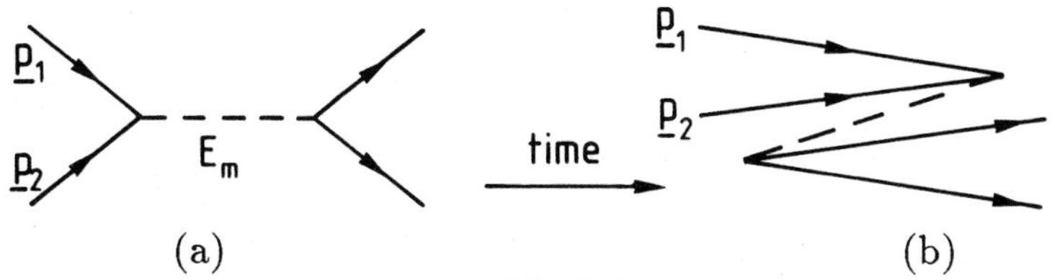


Fig.7.1

The virtual particle carries energy

$$E_m^2 = (\mathbf{p}_1 + \mathbf{p}_2)^2 + m^2 \tag{7.32}$$

in both (a) and (b) of fig.7.1, but with strict time ordering the energy of the intermediate state in (a) is $E_m$, but in (b) is $2E + E_m$, where $E$ is the (well defined) energy in the initial and final states. With the addition of the amplitude of fig.7.1(b) the amplitude for transferring energy $E$ via the field becomes

$$\begin{aligned} \frac{1}{E - E_m} &\rightarrow \frac{1}{E - E_m} + \frac{1}{E - (2E + E_m)} \\ &= \frac{2E_m}{E^2 - E_m^2} \end{aligned} \tag{7.33}$$

and given (7.32)

$$E^2 - E_m^2 = -q^2 - m^2.$$

Comparison of (7.33) with (7.29) and the matrix element

$$\frac{H_{f2}H_{21}}{E - E_2}$$

which is embedded in (7.26) makes it clear that with states normalised to unit probability in unit volume, the amplitude for creating or destroying a meson described by the Klein-Gordon equation must be proportional to $1/\sqrt{E_m}$. This factor vanishes if the normalisation is chosen to be proportional to $E_m$ in unit volume. (Problem 7.2 is relevant.) In either case, the numerator in (7.33) is absorbed into the amplitudes for creating or destroying the meson field and the propagator is our old friend

$$\frac{1}{q^2 + m^2} \tag{7.34}$$

The two amplitudes of fig.7.1 are subsumed into the single diagram of fig.7.2

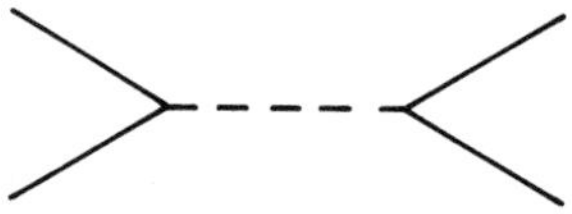

Fig.7.2

where the propagator (7.34) is represented by the broken line. In old fashioned perturbation theory momentum is conserved at each vertex but the intermediate states are off the energy shell. In fig.7.2 both energy and momentum are conserved at each vertex, but the intermediate state is off mass shell; that is, $q^2 + m^2 \neq 0$.

The non-relativistic Breit-Wigner formula, which is often used to represent resonances, is

$$\frac{dP}{dM} = \frac{1}{\pi} \frac{\frac{\Gamma}{2}}{(M_0 - M)^2 + \frac{\Gamma^2}{4}} \tag{7.35}$$

normalised to unity under the assumptions that the resonance is relatively narrow and that $\Gamma$ is approximately independent of mass. The relativistic form is often employed, for reasons which should by now be obvious:

$$\frac{dP}{dM^2} = \frac{1}{\pi} \frac{M_0 \Gamma}{(M_0^2 - M^2)^2 + M_0^2 \Gamma^2} \tag{7.36}$$

where the normalisation is as for (7.35). For hadron resonances, it does not much matter which of the two is used, although fitting (7.35) and (7.36) to the same data for a broad resonance will yield slightly different resonance parameters. It is not even clear that (7.36) is more likely to be correct than (7.35). For fundamental particles and fields the amplitudes represented by figs.7.1(a) and (b) were added without modification because the couplings at the vertices are the same for each. Hadrons have structure and it is not clear that for a massive hadron fig.7.1(b) will not be suppressed by form factors. For fundamental particles and fields the full covariant propagators must be used.

We conclude this section with a few words of warning. First, the propagator in (7.31) is appropriate only for fields fully described by the Klein-Gordon equation—scalar (or pseudoscalar) fields and massless vector fields. Fermions are described by the Dirac equation, which is first order and, in addition to the denominator $q^2 + m^2$, the propagator contains a four-momentum dependent numerator. Massive vector fields also contain a four-momentum dependent numerator. In general the propagator is constructed from the Fourier transform of the fundamental field equations, coupled to a point-like source. Eq.(7.31) is not general.

Secondly, we have not constructed, and shall not construct, the Feynman rules. We have not bothered about the sign of amplitudes, still less about factors of $i$, for the rates which we shall compute are not sensitive to such details. Finally, if the propagator in momentum space is to be Fourier inverted to yield the propagator in configuration space, it is necessary to specify the contour of integration in the vicinity of the poles at $q^2 + m^2 = 0$. This is the reason for the $i\epsilon$ piece to be found in the Feynman propagator, which we shall not need.

### 7.5 Invariant phase space.

In any appliction where relativity is of significance, the Fermi Golden Rule is usually written in a slightly different form. The plane wave in and out states are normalised not to unity in unit volume, but rather to $2E$ in unit volume, where $E$ is the energy of the plane wave state. This choice changes the matrix elements, the flux factor and phase space but of course does not change the physics. The advantage is that the matrix elements are not only invariant, but can often be written easily in a form in which they are manifestly invariant, and the modified phase space is also invariant.

Suppose we seek an expression for probability density. If we have a plane wave confined in a box, the probability that the particle is in the box is unity. If that box is moving with velocity $v$ relative to another observer, at a given instant of time the box is Lorentz contracted, but the integral over the contracted volume of the box must still give unity. The probability density has increased by a factor $(1-v^2)^{-\frac{1}{2}}$—just like energy. More generally, probability must be locally conserved and so there must exist a probability four-current which satisfies the relation

$$\nabla.\mathbf{P} + \frac{\partial P}{\partial t} = 0 \tag{7.37}$$

or

$$\frac{\partial P_\mu}{\partial x_\mu} = 0$$

where $P_\mu = (\mathbf{P}, P)$. The probability density is thus the fourth component of a four-vector and the probability

$$\int P_4 d^3x$$

is an invariant. The only available four-vector is the four-velocity, $dx_\mu/d\tau$. Suppose that in the rest frame of the particle, the wave function is represented by

$$\psi_r = ue^{i(p.x-Et)} \quad (p \to 0)$$

where $u$ may be a number or may contain information about the spin of the particle. Normalise such that $u^+u = 1$; the probability density is unity. If we choose $u^+u$ to be equal to unity in any frame, which is a great convenience, then since

$$p_\mu = mdx_\mu/d\tau$$

the probability density is an arbitrary frame will be

$$P_4 = E/m \qquad \left[P_\mu = \frac{p_\mu}{m}\right]$$

This has obvious disadvantages for massless particles and we normalise such that the probability density of a plane wave is $2m$ in the rest frame of the particle, $2E$ in an arbitrary frame. The volume integral of probability is still an invariant. A plane wave takes the form

$$\psi = \sqrt{2E}ue^{i(p.x-Et)}$$

where $u^+u$ is unity, in all frames. (The factor of 2 is a convenient convention.)

The Golden Rule was worked out in some chosen frame where as a matter of choice the normalisation of plane waves was to unity in unit volume. We rewrite the Golden Rule in the same frame, but with factors of $\sqrt{2E}$ included in the matrix elements. To keep rates the same—remember we are still sitting in the same frame—we must divide by a factor of $2E$ for every particle in the initial state and every particle in the final state. Note that (7.33) already contains a factor $2E_m$ for an intermediate state, which can be absorbed into the matrix elements as normalisation (see problem 7.2), leaving the relativistic Breit-Wigner denominator, or propagator.

When the matrix elements are calculated with plane waves normalised to probability $2E$ in unit volume, the Fermi Golden Rule takes the form

$$dT_{i\to f} = \frac{2\pi}{\prod_i(2E_i)}|\mathcal{M}_{if}|^2 \prod_f^{n-1}\left[\frac{d^3p_f}{(2\pi)^3 2E_f}\right]\frac{1}{2E_n}\frac{1}{dE} \tag{7.38}$$

and the cross section for a process with two incoming particles is obtained by dividing (7.38) by the relative velocity $v$. Remember that the cross section is defined as the rate for a single target exposed to a beam of unit flux: the flux factor $f$ for a normalisation to $2E$ in unit volume is indeed

$$f = \frac{1}{v_{12}2E_1 2E_2} \tag{7.39}$$

The factor

$$dLips = \prod_f^{n-1}\left[\frac{d^3p_f}{(2\pi)^3 2E_f}\right]\frac{1}{2E_n}\frac{1}{dE} \tag{7.40}$$

is an element of Lorentz invariant phase space. The product in (7.40) extends over all independent momentum vectors ($n-1$ for $n$ particles in the final state). The invariance of (7.40) is not manifest, but it can be cast into manifestly invariant form. The flux factor (7.39) is invariant under boosts along the direction of approach of projectile and target, and under such boosts the cross section must be invariant. With the normalisation to $2E$ in unit volume, the matrix elements must be invariants, Lorentz scalars (such as a product of four-scalar products of four-momentum vectors). If the initial state is a single unstable particle, the single factor of $1/2E$ for initial states in (7.38) ensures that for this case the decay rate exhibits the proper time dilation.

The spin 1 fields of QED, QWD and QCD couple to the matrix elements of currents, and these are particularly simply expressed (in terms of four-vector products) using the $2E$ normalisation. If it is necessary to conjecture the form of a matrix element, the knowledge that it must be invariant is invaluable. Finally, a matrix element specified in terms of quantities defined in the centre of mass is by construction an invariant, but not manifestly so. In the decay of hadrons, angular momentum considerations dictate a simple structure for the matrix element in the centre of mass.

We shall not need to employ the so-called relativistic normalisation to $2E$ in unit volume in our subsequent calculations, but its general utility should become apparent in the course of the next three chapters.

## Problems

7.1 If $\psi$ is a solution of the (non-relativistic) free particle Schrödinger equation, it is easy to identify a probability current density $\mathbf{P}$ and a probability density $P$ such that (7.37) is satisfied. If you have never done it, do it now. You will of course find that $P = \psi^*\psi$.

Apply similar manipulations to the relativistic Schrödinger equation, the Klein-Gordon equation, and find $\mathbf{P}$, $P$ such that (7.37) is satisfied. Your results can possibly be identified with probability current and density and multiplied by a charge $e$ can certainly be identified with the electromagnetic current and charge densities.

7.2 While it is not possible to derive QED from Maxwell's equations, we are fortunate that they define a classical field theory which can guide us and to which QED must reduce in the classical limit.

A classical plane wave can be represented by a vector potential

$$\mathbf{A} = a\boldsymbol{\varepsilon}\cos(\mathbf{k}.\mathbf{x} - \omega t)$$

which interacts with a current density $\mathbf{J}$. We want to construct a matrix element for emission or absorbtion of a photon. When the cosine function is broken up into exponentials, we may identify the quantity $a/2$ with the magnitude of the creation and annihilation operators. Calculate the energy density associated with the above plane wave (choose a gauge in which the scalar potential is zero) and set it equal to the energy density for (i) one photon in unit volume (ii) $2\omega$ photons in unit volume. Show that the magnitude associated with the operators is plausibly $\sqrt{4\pi/2\omega}$ in case (i) and merely $\sqrt{4\pi}$ in case (ii).

If you have never done it, show that Maxwell's equations can be written in the form

$$\Box A_\mu = 4\pi J_\mu$$

subject to the gauge condition $\partial_\mu A_\mu = 0$.

7.3* The flux factor, eq.(7.39), is invariant under boosts along the line of flight. Prove it.

Show that the manifestly invariant expression

$$dLips = (2\pi)^3 \delta^{(4)}(\Sigma p_i - \Sigma p_f) \prod_{f=1}^{n} 2\pi\delta(p_f^2 + m_f^2)\frac{d^4 p_f}{(2\pi)^4}$$

can be reduced to eq.(7.40), thereby proving the asserted Lorentz invariance. Note that the above expression contains delta functions for overall conservation of energy and momentum. Proceed by first integrating over the energy components of $p_f$, which are not independent variables because of the mass shell delta functions.

[$\int f(x)\delta(x)dx = f(0)$ if the range of integration includes the point $x = 0$. Work out the value of $\int f(x)\delta(x^2)dx \ldots$]

7.4 Evaluate the Lorentz invariant phase space for a particle of mass $M$ decaying into two other particles, working for convenience in the centre of mass frame, and show that the result is

$$Lips = \frac{1}{8\pi^2}\frac{p}{M}$$

where $p$ is the momentum of one of the two final state particles in the centre of mass.

Suppose that in such a decay the square of the invariant matrix element can be written as a constant, $g^2$. Write down the decay rate in the centre of mass.

Express the invariant phase space in terms of the invariant $s = (\Sigma E_f)^2 - (\Sigma \mathbf{p}_f)^2$ and hence write down an expression for the decay rate which is valid in any frame.

Consider a particle such as the $\rho(770)$, which has $J^P = 1^-$ and decays into two pions in a relative $p$-wave. How would you expect the invariant matrix element, and hence the expression for the decay rate (which is equal to the width term in a Breit-Wigner) to be modified?

7.5* Start from the expression for an element of Lorentz invariant phase space given in Problem 7.3 and consider the case of three particles in the final state. Working in the centre of mass of the three particles, show that the phase space can be written in the form

$$dLips = \frac{1}{64\pi^4} dT_1 dT_2$$

where $T_1$ and $T_2$ are the kinetic energies of two of the three particles, and that this is equivalent to

$$dLips = \frac{1}{256\pi^4 s} dM_{12}^2 dM_{23}^2$$

(which is obviously invariant). This exercise shows that an element of area of the Dalitz plot is proportional to an element of invariant phase space.

For a given total energy, $\sqrt{s}$, calculate $M_{12}^2$ in terms of $M_{23}^2$ and the orientation of particles 2 and 3 relative to the direction of 1, in the rest frame of 2 and 3. This enables you to construct the kinematic boundary of the Dalitz plot.

Express the three particle phase space in terms of the product of invariant two particle phase space for the processes $1 + (2,3)$; $(2,3) \to 2 + 3$.

The $a_1$ ($J^P = 1^{++}$, mass $\sim$ 1260 MeV, width $\sim$ 400 MeV) decays into $\rho\pi$ in a relative $s$-wave. Strong $\rho$ bands appear in the three pion Dalitz plot. What is the distribution of events along the $\rho$ bands? [Ignore the complication of crossing bands; consider a single $\rho - \pi$ pair.]

# 8. FERMIONS AND THE DIRAC EQUATION.

## 8.1 Introduction.

It is impossible to understand much about particle physics without some results from the Dirac equation, which describes elementary fermions. Hard processes in high energy physics, those involving energy or momentum transfers $\gtrsim 10$ GeV, are treated in terms of fermions interacting through vector fields and in many cases first order perturbation theory is sufficient. The fermions are relativistic and intuition based on a knowledge of non-relativistic quantum mechanics is inadequate. The properties of fermions determined by the requirement of a covariant description impinge on the non-perturbative region; we need to understand why the intrinsic parity of fermion and antifermion is negative to see why the pion is $J^P = 0^-$ and why the singlet $s$ state of positronium decays into two photons. After a general introduction we shall set up the Dirac equation in a very simple way.

The essence of the Dirac equation is straightforward. Write down a Lorentz covariant equation that is first order in the time derivative and therefore also in the space derivatives. This can only be done if the four-momentum operators $\hat{\mathbf{p}}$, $i\hat{E}$ are contracted with something else to form a (free particle) equation such as

$$(\gamma_\mu \partial_\mu + im)\psi = 0 \tag{8.1}$$

Since $\partial_\mu$ contains energy and momentum, $\gamma_\mu$ must operate on some internal degree of freedom of $\psi$—the spin degree of freedom. At the same time, $\psi$ must also be a solution of the Klein-Gordon equation in order to preserve

$$E^2 = p^2 + m^2$$

Write

$$H\psi = (\boldsymbol{\alpha}.\hat{\mathbf{p}} + \beta m)\psi \tag{8.2}$$

and require

$$H^2\psi = (\hat{p}^2 + m^2)\psi \tag{8.3}$$

Then for a plane wave solution,

$$(\alpha_i p_i + \beta m)(\alpha_j p_j + \beta m) \equiv p^2 + m^2 \tag{8.4}$$

where we sum over repeated indices. Therefore

$$\begin{aligned} &\alpha_i^2 = 1 \qquad \beta^2 = 1 \\ &\alpha_i\alpha_j + \alpha_j\alpha_i = 2\delta_{ij} \\ &\alpha_i\beta + \beta\alpha_i = 0 \end{aligned} \tag{8.5}$$

and we approach (8.1) with the form

$$(\boldsymbol{\alpha}.\hat{\mathbf{p}} + \beta m)\psi = \hat{E}\psi$$

The anticommutation relations (8.5) for the $\alpha_i$ are just those of the Pauli spin matrices

$$\sigma_x = \begin{pmatrix} 0 & 1 \\ 1 & 0 \end{pmatrix} \qquad \sigma_y = \begin{pmatrix} 0 & -i \\ i & 0 \end{pmatrix} \qquad \sigma_z = \begin{pmatrix} 1 & 0 \\ 0 & -1 \end{pmatrix} \tag{8.6}$$

but $\beta$ cannot be the unit matrix

$$\begin{pmatrix} 1 & 0 \\ 0 & 1 \end{pmatrix}$$

because $\alpha_i\beta + \beta\alpha_i = 0$. Unless $m = 0$ (when we do not need $\beta$) we have to go to $4 \times 4$ matrices, with $\psi$ a 4 component column matrix. The Dirac equation for a free particle is therefore shorthand for four coupled homogeneous equations. For a consistent solution the determinant of the coefficients must vanish, and this (of course) yields the condition $E^2 = p^2 + m^2$. For a given $\mathbf{p}$ there are four degrees of freedom: [positive energy, negative energy (antiparticles)] $\times$ [two spin directions].

The operation Dirac $\longrightarrow$ Klein-Gordon is analogous to Maxwell's equations $\longrightarrow$ Wave equations for **E**, **B** (see Problem 8.5).

### 8.2 Currents and matrix elements.

$\boldsymbol{\alpha}$ and $\beta$ are such that we can write the Dirac equation and its conjugate as

$$\times\psi^+ \qquad (\boldsymbol{\alpha}.\nabla + i\beta m)\psi = -\frac{\partial\psi}{\partial t} \tag{8.7}$$

$$(\nabla.\psi^+\boldsymbol{\alpha} - i\psi^+\beta m) = -\frac{\partial\psi^+}{\partial t} \qquad \times\,\psi \tag{8.8}$$

Multiply (8.7), (8.8) by $\psi^+$, $\psi$ as shown above and add:

$$\nabla.(\psi^+\boldsymbol{\alpha}\psi) + \frac{\partial}{\partial t}(\psi^+\psi) = 0 \tag{8.9}$$

$\psi^+\psi$ is thus the probability density, which is always positive, and $\psi^+\boldsymbol{\alpha}\psi$ is the probability current. The above equation of continuity expresses local conservation of probability. $(\psi^+\boldsymbol{\alpha}\psi,\ \psi^+\psi)$ make up a four-vector and multiplying by electric charge gives the four-current density. These are the pieces which act as the source of the electromagnetic field, $J_\mu$ in (8.10)

$$\nabla^2 A_\mu - \frac{\partial^2}{\partial t^2}A_\mu = -4\pi J_\mu \tag{8.10}$$

(with the choice $\partial_\mu A_\mu = 0$). The Fourier transform of this field, attached to its source, is

$$\tilde{A}_\mu = \frac{4\pi e}{q^2}(u^+\boldsymbol{\alpha}u, u^+u) \tag{8.11}$$

where $\psi = ue^{i(\mathbf{p}.\mathbf{x}-Et)}$ and $u$ is a spinor.

At first sight it is surprising to find $\psi^{+}\boldsymbol{\alpha}\psi$ associated with a (3) current, for you expect a velocity term in there. It is there—for a given energy and spin state $\psi$ has four components, only two of which are independent. It is possible to choose the two dependent terms $\sim {}^{v}\!/_{c}$ relative to components of order 1 and $\boldsymbol{\alpha}$ is off diagonal between these small and large components: $u^{+}\boldsymbol{\alpha}u \sim {}^{v}\!/_{c}$.

The electromagnetic interaction is easily built in. In the equation $H\psi = E\psi$, $E$ is the total energy. In an electric field $H = H_{KE} - e\phi$ for charge $-e$. Then $H_{KE}\psi = (E + e\phi)\psi$ and the free particle equation has been changed by the operation $E \to E + e\phi$

Both $\phi$, $E$ are components of four-vectors, while $e$ is an invariant. To keep the whole equation covariant $p_\mu \to p_\mu + eA_\mu$ and

$$(\boldsymbol{\alpha}.(\hat{\mathbf{p}} + e\mathbf{A}) + \beta m)\psi = (\hat{E} + e\phi)\psi \tag{8.12}$$

The non-relativistic limit of this equation, in a frame where $\phi = 0$ and $\nabla \times \mathbf{A} = \mathbf{B}$ gives energies $\pm\frac{eB}{2m}\left(+\frac{p^2}{2m}\ldots\right)$; that is, the Dirac equation predicts the magnetic moment $\mu = \frac{e\hbar}{2mc}$ (Problem 8.5). Note that $\mathbf{A}$ appears in the form of a scalar product with $\boldsymbol{\alpha}$. Treating $eA_\mu$ as a perturbation on the free field equation, matrix elements for transitions induced by this perturbation are

$$\mathcal{M} \sim e\int \psi_f^{+}\boldsymbol{\alpha}.\mathbf{A}\psi_i \quad , \quad e\int \psi_f^{+}\phi\psi_i \tag{8.13}$$

and so scattering (and annihilation and creation) of charged fermions will be governed by matrix elements

$$\mathcal{M} \sim \frac{4\pi e^2}{q^2}\left\{(u_f^{+}\boldsymbol{\alpha}u_i).(v_f^{+}\boldsymbol{\alpha}v_i), (u_f^{+}u)(v_f^{+}v)\right\} \tag{8.14}$$

where $q^2$ is the square of the four-momentum carried by the field. Eq.(8.14) is usually summarised in the form

$$\mathcal{M} \sim \frac{4\pi e^2}{q^2}(\bar{u}_f\gamma_\mu u_i)(\bar{v}_f\gamma_\mu v_i) \tag{8.15}$$

because it is often convenient to write the Dirac equation in a manifestly covariant form:

$$\begin{aligned}
(\boldsymbol{\alpha}.\nabla + i\beta m)\psi &= -\frac{\partial\psi}{\partial t} \\
(\beta\boldsymbol{\alpha}.\nabla + im)\psi &= -\beta\frac{\partial\psi}{\partial t} && \text{since } \beta^2 = 1 \\
(\gamma_\mu\partial_\mu + im)\psi &= 0 && \boldsymbol{\gamma} = \beta\boldsymbol{\alpha} \quad , \quad \gamma_4 = i\beta
\end{aligned}$$

Then

$$\begin{aligned}
u^{+}\boldsymbol{\alpha}u &\to u^{+}\beta(\beta\boldsymbol{\alpha})u = (-i)\bar{u}\boldsymbol{\gamma}u \\
u^{+}u &\to u^{+}\beta\beta u = (-)\bar{u}\gamma_4 u \qquad \bar{u} = u^{+}\gamma_4
\end{aligned}$$

There are various notations in use—be warned.

### 8.3 The massless Dirac equation.

The easiest way to find an explicit representation of the Dirac equation is to start with the case where $m \to 0$. The general case is then easily constructed, and at high energy, $m \ll E$, the simple properties of the equation without mass still obtain.

For a plane wave and zero mass, the Dirac equation is

$$(E - \boldsymbol{\alpha}.\mathbf{p})\psi = 0$$

and the necessary anticommutation properties of the $\alpha_i$ are those of the three Pauli spin matrices, (8.6). We may choose

$$\boldsymbol{\alpha} = \pm\boldsymbol{\sigma},$$

for

$$(E \pm \boldsymbol{\sigma}.\mathbf{p})\psi = 0 \tag{8.16}$$

satisfies

$$E^2 - p^2 = 0.$$

The massless fermion is represented by

$$\psi = \begin{pmatrix} u_1 \\ u_2 \end{pmatrix} e^{i(\mathbf{p}.\mathbf{x} - Et)}$$

and taking for the positive sign in (8.16) $\psi(+) = \chi$

$$(E + \boldsymbol{\sigma}.\mathbf{p})\chi = (E + \boldsymbol{\sigma}.\mathbf{p})\begin{pmatrix} u_1 \\ u_2 \end{pmatrix} = 0 \tag{8.17}$$

Expand into two coupled equations:

$$\begin{aligned} Eu_1 + (p_x + ip_y)u_2 + p_z u_1 &= 0 \\ Eu_2 + (p_x - ip_y)u_1 - p_z u_2 &= 0 \end{aligned} \tag{8.18}$$

These are consistent if the determinant of the coefficient matrix vanishes — $E^2 - p^2 = 0$.
The two equations are decoupled by choosing the $z$ axis to lie along $\mathbf{p}$:

$$\begin{aligned} (E + p)u_1 &= 0 \\ (E - p)u_2 &= 0 \end{aligned} \tag{8.19}$$

Then $\binom{1}{0}$ is a positive eigenstate of $\boldsymbol{\sigma}.\mathbf{p}/p$ and has negative energy ($E = -p$)
$\binom{0}{1}$ is a negative eigenstate of $\boldsymbol{\sigma}.\mathbf{p}/p$ and has positive energy ($E = +p$)
Eigenstates of $\boldsymbol{\sigma}.\mathbf{p}/p$ are helicity states, so $(E + \boldsymbol{\sigma}.\mathbf{p})\chi = 0$ describes a particle with left helicity (which is also left handed for $m \to 0$) and an anti-particle with right helicity (right handed for $m \to 0$)

Fig.8.1

The entirely separate equation, for $\psi(-) = \phi$

$$(E - \boldsymbol{\sigma}.\mathbf{p})\phi = 0 \tag{8.20}$$

describes a right helicity particle and a left helicity antiparticle. Since these (massless) equations are Lorentz covariant (we have not explicitly proved it) the solutions have this form in any inertial frame. Admittedly $\mathbf{p}$ changes direction (as well as magnitude) under a Lorentz boost, but so does $\boldsymbol{\sigma}$—in such a way that a (massless) left handed particle is always lefthanded. The Lorentz transformation properties of spin $1/2$ emerge from the covariant Dirac equation.

Electromagnetic interactions enter through $e\boldsymbol{\alpha}.\mathbf{A}$, $e\phi$ and for these decoupled equations (8.17, 8.20), which have solutions $\chi$, $\phi$ respectively, the source of $A_\mu$ will be $(\chi^+\boldsymbol{\sigma}\chi,\ \chi^+\chi)$ or $(\phi^+\boldsymbol{\sigma}\phi,\ \phi^+\phi)$, and at the other end perturbations will be $\sim \chi^+\boldsymbol{\sigma}.\mathbf{A}\chi$ etc. Any field with four-vector properties under rotations and boosts (QED, QWD, QCD) thus couples only

| | | | |
|---|---|---|---|
| left handed particle | to | left handed particle | |
| right handed particle | to | right handed particle ... | (8.21) |
| left handed particle | to | right handed antiparticle | |
| right handed particle | to | left handed antiparticle | |

This is enormously useful in understanding the characteristics of these interactions†. These results are not intuitively obvious.

The charge changing weak interactions are even more specific. They only recognise the existence of left handed particles and right handed antiparticles. This is where non-conservation of parity in the weak interactions is expressed, because under parity $(\mathbf{x} \to -\mathbf{x})$ $\mathbf{p} \to -\mathbf{p}$ but $\boldsymbol{\sigma} \to \boldsymbol{\sigma}$. The parity operation may be represented by a mirror, as in fig.8.2

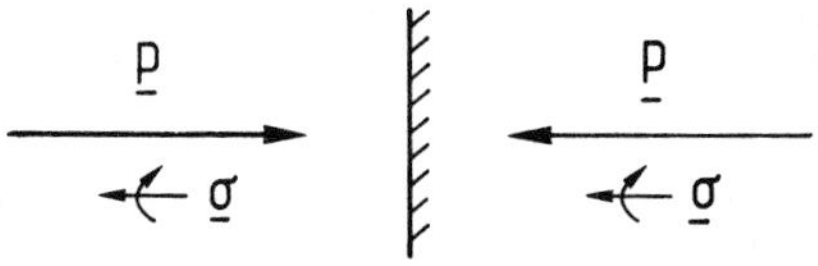


Left handed particle    Right handed particle—no coupling to $W^\pm$

Fig.8.2

[Even though the (massless) Hamiltonian $\pm\boldsymbol{\sigma}.\mathbf{p}$ is ODD, in strong and electromagnetic interactions parity is not violated, because left handed and right handed particles enter on exactly the same footing:

† In this limit there is no distinction between helicity and handedness. The distinction is introduced in section 8.5.

Fig.8.3

]

### 8.4 Rotational properties of spin.

Look at the solutions of the massless equations for the slightly more general case $p_y = 0$, $p_x \neq 0$. For the $\chi$ set (left handed particles) $\chi = \binom{u_1}{u_2}$ and

$$\begin{aligned} Eu_1 + p_x u_2 + p_z u_1 &= 0 \\ Eu_2 + p_x u_1 - p_z u_2 &= 0 \end{aligned} \tag{8.22}$$

For positive energy

$$u_1 = -\frac{p_x u_2}{E + p_z}$$

Then

$$\chi = N\begin{pmatrix} \frac{-p_x}{E+p_z} \\ 1 \end{pmatrix} \qquad \left[\text{and } \chi \to \begin{pmatrix} 0 \\ 1 \end{pmatrix} \text{ as } p_x \to 0\right]$$

$$\chi^+\chi = N^2\left\{1 + \frac{p_x^2}{(E+p_z)^2}\right\}$$

Set $p_x = p\sin\theta$, $p_z = p\cos\theta$ and remembering that $E = p$ for the massless case, we obtain

$$\chi^+\chi = N^2\left\{1 + \frac{\sin^2\theta}{(1+\cos\theta)^2}\right\} \quad ; \quad \chi = N\begin{pmatrix} \frac{-\sin\theta}{1+\cos\theta} \\ 1 \end{pmatrix}$$

$$\chi^+\chi = \frac{2N^2}{1+\cos\theta} = \frac{N^2}{\cos\frac{\theta}{2}} = 2E$$

using the conventional normalisation. Then

$$\begin{aligned} \chi &= \sqrt{2E}\cos\frac{\theta}{2}\begin{pmatrix} -2\sin\frac{\theta}{2}\cos\frac{\theta}{2} / 2\cos^2\frac{\theta}{2} \\ 1 \end{pmatrix} \\ &= \sqrt{2E}\begin{pmatrix} -\sin\frac{\theta}{2} \\ \cos\frac{\theta}{2} \end{pmatrix} \quad \to \quad \sqrt{2E}\begin{pmatrix} 0 \\ 1 \end{pmatrix} \text{ as } \theta \to 0 \end{aligned} \tag{8.23}$$

You can see that it might be advantageous to keep the spinor parts normalised to unity in any frame. The expressions (8.23) still represent the left handed particle solution but now referred to external axes with $\mathbf{p}$ in the $x, z$ plane

$$\begin{pmatrix} -\sin\frac{\theta}{2} \\ \cos\frac{\theta}{2} \end{pmatrix} = -\sin\frac{\theta}{2}\begin{pmatrix} 1 \\ 0 \end{pmatrix} + \cos\frac{\theta}{2}\begin{pmatrix} 0 \\ 1 \end{pmatrix} \tag{8.24}$$

and this is how the left handed state projects into spin components measured with respect to a $z$ axis an angle $\theta$ away from the momentum vector. It corresponds to the well known rotation properties of spin $1/2$ and the result is going to be important in understanding the properties of fermion-fermion interactions. [The rotational properties of spin $1/2$ are curious. You can see that under a rotation through 360° the wave function does not return to its original value, but picks up a phase change of $\pi$. This has been experimentally verified with a neutron interferometer. A very interesting discussion of the antisymmetry of spin $1/2$ wave functions has been given in terms of these rotational properties: R.P. Feynman in *Elementary Particles and the Laws of Physics* Cambridge University Press (1987).] You can also see that in this limit $m/E \to 0$ it will be rather simple to calculate explicitly matrix elements involving spin 1 fields. Anyone can multiply $2 \times 2$ matrices. [$4 \times 4$ is another matter ... you need to learn the tricks.]

### 8.5 The reinsertion of mass.

The two decoupled Weyl equations, (8.17) and (8.20), violate parity if considered separately. The left handed equation, (8.17), can be used to describe neutrinos if they really have zero mass. We have asserted that in the high energy limit the Dirac equation decouples in this way; to justify this assertion we first construct the Dirac equation with mass. We start by assembling the two massless equations into a single equation with four component spinor solutions:

$$\left.\begin{matrix}(E+\boldsymbol{\sigma}.\mathbf{p})\chi = 0\\ (E-\boldsymbol{\sigma}.\mathbf{p})\phi = 0\end{matrix}\right\}(E-\boldsymbol{\alpha}.\mathbf{p})\psi = 0 \quad ; \quad \psi = \binom{\chi}{\phi} \tag{8.25}$$

We make the identification $\boldsymbol{\alpha} = \begin{pmatrix} -\boldsymbol{\sigma} & 0 \\ 0 & \boldsymbol{\sigma} \end{pmatrix}$ which is of course diagonal in $\chi, \phi$. In order to insert a mass, we must introduce a matrix $\beta$ which squares to give 1 and satisfies $\alpha\beta + \beta\alpha = 0$. This is easy —

$$\beta = \begin{pmatrix} 0 & 1 \\ 1 & 0 \end{pmatrix}$$

This matrix is off diagonal between $\chi, \phi$ so a mass term yields eigenstates which contain both, four components in all. We also have

$$\beta\boldsymbol{\alpha} = \boldsymbol{\gamma} = \begin{pmatrix} 0 & \boldsymbol{\sigma} \\ -\boldsymbol{\sigma} & 0 \end{pmatrix}; \quad \gamma_4 = i\beta = i\begin{pmatrix} 0 & 1 \\ 1 & 0 \end{pmatrix}$$

We now have a representation (the Weyl representation) of the Dirac spin matrices. [It is not unique: the representation usually encountered is different, the Pauli-Dirac representation.]

Now

$$\begin{pmatrix} 1 & 0 \\ 0 & 0 \end{pmatrix}\begin{pmatrix} \chi \\ \phi \end{pmatrix} = \begin{pmatrix} \chi \\ 0 \end{pmatrix} \qquad \left[\begin{pmatrix} 0 & 0 \\ 0 & 1 \end{pmatrix}\begin{pmatrix} \chi \\ \phi \end{pmatrix} = \begin{pmatrix} 0 \\ \phi \end{pmatrix}\right] \tag{8.26}$$

so

$$\begin{pmatrix} 1 & 0 \\ 0 & 0 \end{pmatrix}$$

projects out left handed particles ($\chi$) and right handed antiparticles. This can be written

$$\begin{pmatrix} 1 & 0 \\ 0 & 0 \end{pmatrix} = \frac{1}{2}(1-\gamma_5) \quad ; \quad \gamma_5 = \begin{pmatrix} -1 & 0 \\ 0 & 1 \end{pmatrix} = -\gamma_1\gamma_2\gamma_3\gamma_4 \tag{8.27}$$

The charged (which means charge changing) weak current is

$$J_{weak\ \mu} = \bar{\psi}\gamma_\mu \frac{1}{2}(1-\gamma_5)\psi \tag{8.28}$$

and $\frac{1}{2}(1-\gamma_5)$ projects out the $\chi$ (left handed) states from a state which is a mixture of $\chi$ and $\phi$ if $m$ is not zero. Handedness is not a good quantum number if $m \neq 0$.

| | | | |
|---|---|---|---|
| $\bar{\psi}\gamma_\mu\psi$ | is a | (4-) Vector | (8.29($a$)) |
| $\bar{\psi}\gamma_\mu\gamma_5\psi$ | is an | Axial Vector. | (8.29($b$)) |

The four-vector nature of (8.29(a)) should be clear from the identification of the conserved probability current, (8.9). The axial vector (8.29(b)) has four-vector properties under rotations and boosts, but does not change sign under parity. [Both angular momentum and magnetic field are axial vectors.]

The weak interaction has $V-A$ form. Note that $(1-\gamma_5)(1-\gamma_5) = 2(1-\gamma_5)$ and $\gamma_\mu\gamma_5 = -\gamma_5\gamma_\mu$ so

$$J_{weak\ \mu} \sim \psi^+(1-\gamma_5)\gamma_4\gamma_\mu(1-\gamma_5)\psi \tag{8.30}$$

and $(\chi\phi)\begin{pmatrix} 1 & 0 \\ 0 & 0 \end{pmatrix} = (\chi\ 0)$ so one factor of $(1-\gamma_5)$ projects left handed leptons on both sides.

Handed (or chiral) states $\binom{\chi}{0}$, $\binom{0}{\phi}$ are projected out by $(1 \pm \gamma_5)$ and are coupled by vector and axial vector interactions, according to the rules given in (8.21). They are not helicity states except in the limit $m/E \to 0$. Helicity states are eigenstates of the operator $\mathbf{s}.\mathbf{p}/p$ (where $\mathbf{s} = \begin{pmatrix} \boldsymbol{\sigma} & 0 \\ 0 & \boldsymbol{\sigma} \end{pmatrix}$) and in the Weyl representation are easily found from the complete (4 component) Dirac equation by setting $p_z = p$ ($p_x = p_y = 0$). They are

Positive energy — Negative energy

$$\underset{\text{Left}}{\begin{pmatrix} 0 \\ 1 \\ 0 \\ K \end{pmatrix}}, \underset{\text{Right}}{\begin{pmatrix} K \\ 0 \\ 1 \\ 0 \end{pmatrix}} \qquad \underset{\text{Left}}{\begin{pmatrix} 0 \\ -K \\ 0 \\ 1 \end{pmatrix}}, \underset{\text{Right}}{\begin{pmatrix} 1 \\ 0 \\ -K \\ 0 \end{pmatrix}} \quad \text{HELICITY} \tag{8.31}$$

where $K = \frac{m}{|E|+p} = \frac{\sqrt{1-\beta}}{\sqrt{1+\beta}}$. As $\beta(= {}^{v}/_{c}) \to 1$, $K \to 0$ and the handed states result in this limit. The right helicity state contains a left handed piece, in general

$$\left[\tfrac{1}{2}(1-\gamma_5) = \begin{pmatrix} 1 & 0 \\ 0 & 0 \end{pmatrix}\right] \begin{pmatrix} K \\ 0 \\ 1 \\ 0 \end{pmatrix} = \begin{pmatrix} K \\ 0 \\ 0 \\ 0 \end{pmatrix} = K \begin{pmatrix} 1 \\ 0 \\ 0 \\ 0 \end{pmatrix} \tag{8.32}$$

Left hand projection Right helicity

In pion decay the diagram is that of fig.8.4(a)

Fig.8.4

In the centre of mass the $z$ components of lepton spin sum to zero, fig.8.4(b), so the helicities are the same. But the $W^+$ couples to the left handed neutrino and the right handed component of the $\mu^+$. Thus there is a factor $\sim \sqrt{1-\beta_\mu}$ in the matrix element, because this is the amount of left handed $\mu^+$ in the right helicity state. In the limit where both lepton masses$\to 0$, the pion could not decay. As it is, $1-\beta_\mu \gg 1-\beta_e$ and this is why $\pi^+ \to e^+\nu_e/\pi^+ \to \mu^+\nu_\mu \simeq 10^{-4}$ (Problem 8.1).

### 8.6 Fermions, antifermions and parity.

We need one more result from the Dirac equation; that the relative parity of fermion and antifermion is negative. We can only extract this result from something which describes fermion and antifermion together in a single package: the Dirac equation with both positive and negative energy solutions.

The parity operation is inversion of spatial coordinates $\mathbf{x} \to -\mathbf{x}$. Denoting the new coordinates by $\mathbf{x}'$, $x_i' = -x_i$. Under this operation, vectors change sign but axial vectors, defined through cross products, do not. A quantity such as $\boldsymbol{\sigma}.\mathbf{p}$ is a pseudoscalar and changes sign. Coordinate inversion does not affect the matrices $\boldsymbol{\alpha}$ and $\beta$, which contain coordinates nowhere. We thus expect

$$P\left\{(\hat{E} - \boldsymbol{\alpha}.\hat{\mathbf{p}} - \beta m)\psi\right\} = (\hat{E} + \boldsymbol{\alpha}.\hat{\mathbf{p}} - \beta m)\psi_P \tag{8.33}$$

where

$$\psi_P = P\psi$$

Because $\boldsymbol{\alpha}$ and $\beta$ anticommute

$$\beta\left\{(\hat{E} - \boldsymbol{\alpha}.\hat{\mathbf{p}} - \beta m)\psi\right\} = (\hat{E} + \boldsymbol{\alpha}.\hat{\mathbf{p}} - \beta m)\beta\psi \tag{8.34}$$

and without more ado we identify $\psi_P$ with $\beta\psi$. The above argument is correct but deceptively simple. It is worth spelling it out in more detail.

$$P(x_i) = x_i' \quad ; \quad x_i' = -x_i$$

$$P\left\{(\hat{E}-\boldsymbol{\alpha}.\hat{\mathbf{p}}-\beta m)\psi(\mathbf{x})\right\}=(\hat{E}-\boldsymbol{\alpha}.\hat{\mathbf{p}}'-\beta m)\psi'(\mathbf{x}') \tag{8.35}$$

We require, for a plane wave,

$$\frac{1}{i}\frac{\partial}{\partial x_i'}\psi'(x_i')=p_i'\psi'(x_i') \tag{8.36}$$

and $p_i'=-p_i$. The phase factor in $\psi'(x_i')$ must then be

$$\psi'(x_i')\sim e^{+ip_i'x_i'}=e^{-ip_ix_i'}=e^{ip_ix_i} \tag{8.37}$$

The phase factor in $\psi'(x_i')$ can be written as the phase factor in $\psi(x_i)$ and so it is legitimate to seek a parity operator which only involves the Dirac matrices. If we rewrite $\mathbf{p}'$ in terms of $\mathbf{p}$, then from (8.35)

$$P\left\{(\hat{E}-\boldsymbol{\alpha}.\hat{\mathbf{p}}-\beta m)u\right\}=(\hat{E}+\boldsymbol{\alpha}.\hat{\mathbf{p}}-\beta m)u_P \tag{8.38}$$

and $\beta u$ may be identified with $u_P$.

We have written $u_P$ in such a form that it is a solution of the original Dirac equation, with the sign of $\boldsymbol{\alpha}.\hat{\mathbf{p}}$ changed. This simply corresponds to interchanging the components $\binom{\chi}{\phi}$ of the spinors.

$$P\begin{pmatrix}\chi\\ \phi\end{pmatrix}=\beta\begin{pmatrix}\chi\\ \phi\end{pmatrix}=\begin{pmatrix}\phi\\ \chi\end{pmatrix} \tag{8.39}$$

If this operation is applied to the positive energy left helicity solution in (8.31)

$$P\begin{pmatrix}0\\1\\0\\K\end{pmatrix}\rightarrow\begin{pmatrix}0\\K\\0\\1\end{pmatrix} \tag{8.40}$$

and this is a right helicity solution in the coordinates $x_i'=P(x_i)=-x_i$, for the helicity operator is now $-\sigma_z$; we have defined our solutions with respect to a $z$ axis along $\mathbf{p}$.

The helicity states are obviously not eigenstates of parity, and in general plane wave solutions of the Dirac equation are not eigenstates of parity except in the limit $E\rightarrow m$, $p\rightarrow 0$. It suffices to consider only this limit in order to extract the relative parity of fermion and antifermion. From (8.39)

$$P\begin{pmatrix}\chi\\ \phi\end{pmatrix}=\begin{pmatrix}\phi\\ \chi\end{pmatrix}$$

and in the limit $p\rightarrow 0$ (8.31) yields

$$\begin{aligned}\phi&=\chi &&\text{for positive energy, and}\\ \phi&=-\chi &&\text{for negative energy.}\end{aligned}$$

Thus

$$
\begin{aligned}
Pu^{+}(\mathbf{p}=0) &= +u^{+}(\mathbf{p}=0) \\
Pu^{-}(\mathbf{p}=0) &= -u^{-}(\mathbf{p}=0)
\end{aligned}
$$

The positive and negative energy solutions, particle and antiparticle, have opposite intrinsic parity. [With the identification of $\beta$ (or $\gamma_4$) with the parity operator, it becomes obvious that $\boldsymbol{\gamma}$ is a vector operator, $\boldsymbol{\gamma}\gamma_5$ is an axial vector operator and $\gamma_5$ is a pseudoscalar operator. We have $\beta\gamma_5 = -\gamma_5\beta$, $\beta[\boldsymbol{\gamma}\gamma_5] = [\boldsymbol{\gamma}\gamma_5]\beta \ldots$]

We may conclude that the $s$-wave states of positronium, $e^+e^-$, have negative parity, and finally understand how the negative parity of the pion originates. The pion is a spin zero $s$-wave state of quark and antiquark. The quark spins are not changed by the parity operation, the spatial part of the wavefunction is even under parity and the negative parity of the pion has its origin in the negative intrinsic parity of a fermion-antifermion pair.

We may extract finally the properties of a state consisting of a fermion-antifermion pair under the operation of charge conjugation, $C$. This operation turns a particle into an antiparticle—that is all. Thus $C(\nu_L) \to \bar{\nu}_L$ and nature is not invariant under $C$ because $\bar{\nu}_L$ does not exist. [The statement is still true even if $\nu_R$ do exist, because the weak couplings to a possible $\nu_R$ are negligible in comparison with the coupling to $\nu_L$.]

Bound states of a fermion and its antiparticle are eigenstates of charge conjugation

$$
C(f\bar{f}) = (\bar{f}f) = \pm(f\bar{f})
$$

Thus the $\pi^\circ$ (composed of $u\bar{u}$, $d\bar{d}$) is an eigenstate of $C$ but $\pi^+$ ($u\bar{d}$) is not.

Consider the states

$$
\underset{1}{\bar{\uparrow}}\underset{2}{\downarrow} \pm \underset{1}{\bar{\downarrow}}\underset{2}{\uparrow}
$$

where the arrows refer to spins and the labels 1, 2 to coordinates. We represent

$$
\begin{aligned}
C\,(\underset{1}{\bar{\uparrow}}\underset{2}{\downarrow} \pm \underset{1}{\bar{\downarrow}}\underset{2}{\uparrow}) &= \underset{1}{\uparrow}\underset{2}{\bar{\downarrow}} \pm \underset{1}{\downarrow}\underset{2}{\bar{\uparrow}} \\
S\,(\bar{\uparrow}\downarrow \pm \bar{\downarrow}\uparrow) &= \bar{\downarrow}\uparrow \pm \bar{\uparrow}\downarrow
\end{aligned}
$$

where $S$ is the spin interchange operation. Finally

$$
P\,(\bar{\downarrow}\uparrow \pm \bar{\uparrow}\downarrow) = (\uparrow\bar{\downarrow} \pm \downarrow\bar{\uparrow})
$$

The operations $PS$ yield the same state as $C$. We know that

$$
S\psi = (-1)^{S+1}\psi
$$

(positive eigenvalue for the symmetric spin triplet; negative value for the antisymmetric spin singlet) and for fermion-antifermion

$$
P\psi = (-1)(-1)^{L}\psi
$$

where the first factor comes from the negative intrinsic parity of the fermion-antifermion pair. Thus we may identify

$$C\psi(f\bar{f}) = (-1)^{L+S}\psi(f\bar{f}) \tag{8.41}$$

and any hadron which may be represented as an $L-S$ decomposition of quark-antiquark must satisfy

$$C = (-1)^{L+S}$$

Thus a $q\bar{q}$ state with $L = 0$, $S = 0$ has negative parity and positive $C$; $L = 0$, $S = 1$ has negative parity and negative $C$. The $\pi^\circ$ is a state $J^{PC} = 0^{-+}$ and decays into $\gamma\gamma$ (a $C = +1$ final state). The ground state of positronium is $L = 0$, $S = 0$ and decays into $\gamma\gamma$; therefore $C = +1$. The polarisation vectors $\boldsymbol{\varepsilon}_1$, $\boldsymbol{\varepsilon}_2$ of the two photons are distributed according to $[\mathbf{k}.(\boldsymbol{\varepsilon}_1 \times \boldsymbol{\varepsilon}_2)]^2$, where $\mathbf{k}$ is the relative momentum vector. The state represented by $\mathbf{k}.(\boldsymbol{\varepsilon}_1 \times \boldsymbol{\varepsilon}_2)$ is a pseudoscalar, odd under parity. The ground state of positronium (which is undoubtedly a fermion-antifermion bound state) is found experimentally to have $C = +1$ and negative parity in accord with the Dirac equation. The triplet state cannot decay into one real photon, but decays into three, a $C = -1$ final state.

## Problems

8.1 At the end of section 8.5 we asserted that the branching ratio for $\pi^+ \to e^+\nu_e$ is $\sim 10^{-4}$. Show that

$$\frac{\Gamma(\pi^+ \to e^+\nu_e)}{\Gamma(\pi^+ \to \mu^+\nu_\mu)} = \left(\frac{m_e}{m_\mu}\right)^2 \left(\frac{m_\pi^2 - m_e^2}{m_\pi^2 - m_\mu^2}\right)$$

[Assume $m_{\nu_e} = m_{\nu_\mu} = 0$]. Evaluate this expression, and make a reasoned estimate of the lifetime of the charged pion.

Evaluate the branching ratio $K^+ \to e^+\nu_e/K^+ \to \mu^+\nu_\mu$ (Experimental values: $\pi^+ \to e^+\nu_e/\pi^+ \to \mu^+\nu_\mu = 1.23 \times 10^{-4}$; $K^+ \to e^+\nu_e/K^+ \to \mu^+\nu_\mu = 2.42 \times 10^{-5}$).

8.2 The $W^\pm$ bosons which mediate the weak interactions have spin 1 and mass $\sim$ 82 GeV. They couple only to left handed leptons and quarks, right handed anti-leptons and anti-quarks. They were first observed with the CERN $Sp\bar{p}S$ collider in which beams of protons and anti-protons circulated in opposite directions with momenta of 270 GeV. The signature consisted of the observation of a muon (or electron) with high energy at a large angle to the beams, with approximately equal and opposite missing transverse energy.

Show that the angular distribution of leptons ($e^-$, $\mu^-$) from $W^-$ production is

$$\sim (1 + \cos\theta)^2$$

relative to the proton direction, while the angular distribution of antileptons ($e^+$, $\mu^+$) from $W^+$ production

$$\sim (1 - \cos\theta)^2$$

relative to the proton direction. [This is an important check that $W^{\pm}$ are the sources of the high transverse energy leptons. In the early days there were some queasy moments before sufficient statistics accumulated for the verification of these predictions.]

8.3 Begin by verifying the relations (8.31). Normalise these four component spinors and demonstrate that

$$u^{+}\boldsymbol{\alpha}u = \hat{\mathbf{k}}\frac{1-K^2}{1+K^2} = \mathbf{v}$$

where $\hat{\mathbf{k}}$ is a unit vector in the direction of the momentum (and units are, as usual, $c = 1$). [In the Weyl representation the dependent component is small as $v \to c$. In the Pauli-Dirac representation the dependent component is small as $v \to 0$. In both cases $u^{+}\boldsymbol{\alpha}u = \mathbf{v}$.]

Consider the case of the positive energy, left helicity solution of the Dirac equation in the Weyl representation. The first column of (8.31) is obtained by setting $p_z = p$; $p_x = p_y = 0$. Choose the $z$ axis at some angle $\theta$ such that $p_z = p\cos\theta$, $p_x = p\sin\theta$ and show that the positive energy left helicity spinor, normalised to unity, is

$$\frac{1}{\sqrt{1+K^2}}\begin{pmatrix} -\sin\frac{\theta}{2} \\ \cos\frac{\theta}{2} \\ -K\sin\frac{\theta}{2} \\ K\cos\frac{\theta}{2} \end{pmatrix}$$

8.4* The Weyl representation of the Dirac matrices has obvious advantages for high energy physics. In the Pauli-Dirac representation the spin eigenstates have only one non-zero component in the low velocity limit and the formalism reduces to that of two component Pauli spinors. Nonetheless, the Weyl representation can be used to advantage to understand some features of nuclear $\beta$ decay. You are invited to study the angular correlation between electron and neutrino obtaining in ${}^{14}\mathrm{O} \to {}^{14}\mathrm{N}^{*}e^{+}\nu_e$ and in neutron $\beta$ decay $n \to pe^{-}\bar{\nu}_e$. The results are easily obtained without any of the apparatus of taking the traces of the products of $\gamma$ matrices. The weak interaction for $\beta$ decay is proportional to

$$[\bar{\psi}\gamma_\mu(1-\gamma_5)\psi]\left[\bar{\Psi}\gamma_\mu(1-\frac{g_A}{g_V}\gamma_5)\Psi\right]$$

where the first bracket contains the lepton fields and the second the nucleon wave functions. [The ratio $g_A/g_V$ is believed to be unity for the quarks. For the decay of the free neutron it has a value 1.25. There is evidence that in complex nuclei this effect of quark binding is attenuated.]

${}^{14}\mathrm{O}$ and ${}^{14}\mathrm{N}^{*}$ are both $0^{+}$ members of a nuclear isospin triplet. The transition is super-allowed pure Fermi. Show that the electron-neutrino angular correlation is

$$1 + \frac{v_e}{c}\cos\theta_{e\nu}$$

Does the presence of the factor $1-\gamma_5$ in the lepton bracket have any influence on this result?

Show that the angular correlation in neutron decay is given by

$$\frac{1}{4}\left\{1+3\left(\frac{g_A}{g_V}\right)^2+\left(1-\left(\frac{g_A}{g_V}\right)^2\right)\frac{v_e}{c}\cos\theta_{e\nu}\right\}$$

[Treat the nucleons as non-relativistic: only $\gamma_4$ survives in the nucleon bracket. For simplicity take the $z$ axis along the direction of the electron and refer the neutrino spinors to this direction. In order to get the sign of the angular correlation right, be particularly careful about the sign of the momentum to be associated with the antiparticle solutions.]

Show that electrons are longitudinally polarised to degree $v_e/c$. [The longitudinal polarisation of neutrinos is obvious.]

8.5* In the Weyl representation, the free particle Dirac equation can be written in the form

$$\boldsymbol{\sigma}.\boldsymbol{\nabla}\chi=\frac{\partial\chi}{\partial t}+im\phi$$
$$\boldsymbol{\sigma}.\boldsymbol{\nabla}\phi=-\frac{\partial\phi}{\partial t}-im\chi$$

where the operation $\boldsymbol{\sigma}.\boldsymbol{\nabla}$ is analogous to the vector operator *curl*, in that $(\boldsymbol{\sigma}.\boldsymbol{\nabla})(\boldsymbol{\sigma}.\boldsymbol{\nabla})=\nabla^2$. Prove this relation and eliminate the Pauli spinors $\chi$, $\phi$ in turn to show that each satisfies the Klein-Gordon equation. The operation Dirac→Klein-Gordon is analogous to the reduction of Maxwell's equations to wave equations for the electromagnetic fields. The analogy becomes closer if the Dirac matrices are chosen so that $\beta$ is diagonal, and the two-spinor equations are coupled through the time derivative. Show that the equations

$$\boldsymbol{\sigma}.\boldsymbol{\nabla}\eta=\frac{\partial\rho}{\partial t}-im\rho$$
$$\boldsymbol{\sigma}.\boldsymbol{\nabla}\rho=\frac{\partial\eta}{\partial t}+im\eta$$

also yield a pair of Klein-Gordon equations, and hence that a possible choice for $\boldsymbol{\alpha}$ and $\beta$ is

$$\boldsymbol{\alpha}=\begin{pmatrix}0&\boldsymbol{\sigma}\\\boldsymbol{\sigma}&0\end{pmatrix}\quad;\quad\beta=\begin{pmatrix}1&0\\0&-1\end{pmatrix}$$

This is the Pauli-Dirac representation. Show that in the low velocity limit $v\to 0$ the eigenstates of $s_z$ have only one of the four spinor components non-zero.

Working in this representation, consider the Dirac equation in a frame in which there exists a magnetic field $\mathbf{B}=\boldsymbol{\nabla}\times\mathbf{A}$ in the $z$ direction. Extract the non-relativistic Schrödinger equation limit for the large components and show that the eigenstates of $s_z$ have energy $\pm\frac{eB}{2m}$. [A pointlike fermion has magnetic momentum $\frac{e\hbar}{2mc}$.]

An interesting extension is to consider the Dirac equation in a frame in which a potential exists. The spin-orbit coupling can be extracted by finding the Schrödinger equation for the large components, eliminating the small. The potential may be either the fourth component of a four-vector, as in an atom, or a Lorentz scalar. Make sure that you start from a manifestly covariant form of the Dirac equation. You will find that the signs of the spin-orbit coupling are opposite for the fourth component of four-vector potentials and a (Lorentz) scalar. [Remember that the strong spin-orbit coupling in nuclei is opposite to that obtaining in atoms. Spin-orbit coupling in hadrons seems to be rather weak, suggesting that the long-range quark confining potential is scalar in character.]

## 9. FERMION-FERMION INTERACTIONS.

### 9.1 $e^+e^- \to \mu^+\mu^-\ldots$

While the calculation of, for example, $e^+e^- \to \mu^+\mu^-$ via one photon annihilation

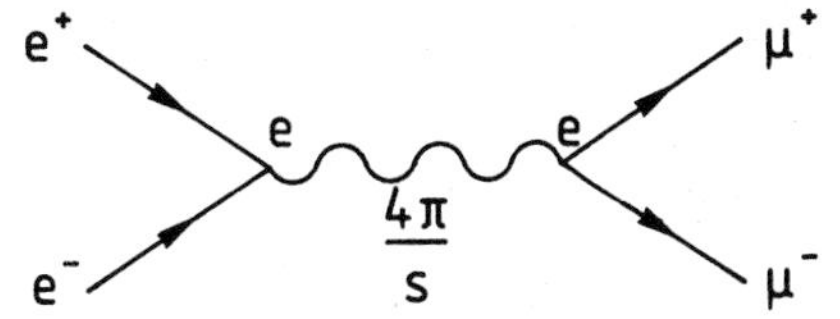


Fig.9.1

involves a deal of algebra if done in all generality, in the limit where fermion masses may be neglected it can be performed rather easily. First, it is obvious on dimensional grounds that in this limit

$$\sigma \sim e^4/s$$

[in natural units; $s$ is the square of the centre of mass energy.]

We might hope to get this by crossing Coulomb scattering from space-like (momentum carrying) to time-like (energy carrying) photons, but this will not do. The current-current interactions must be included as well as the charge-charge interaction. The appropriate Fourier components of the electromagnetic field with a source $j_\mu$ are

$$\tilde{A}_\mu \sim 4\pi\tilde{j}_\mu/q^2 \tag{9.1}$$

where

$$j_\mu \sim e\bar{\psi}\gamma_\mu\psi$$

and

$$\tilde{j}_\mu \sim e\bar{u}\gamma_\mu u$$

At the other end, $A_\mu$ couples to $j'_\mu$ and we have a matrix element

$$\mathcal{M} \sim \frac{e^2(\bar{u}'\gamma_\mu u')(\bar{u}\gamma_\mu u)}{q^2} \sim \frac{e^2}{q^2} \quad ; \quad |\mathcal{M}|^2 \sim \frac{e^4}{s^2} \tag{9.2}$$

The cross section does not vary as $1/s^2$ because of phase space

$$\rho \sim p^2 dp/dE_{\rm cm}.$$

In the high energy limit $E_{\rm cm} = 2p$ and $p^2 = s/4$ so that

$$\sigma \sim e^4/s$$

results.

In the above expressions the spinors were normalised to unity and non-relativistic phase space was used. We get the same result using Lorentz invariant phase space and normalising $\psi^+\psi = 2E$. In the centre of mass four factors of

$\sqrt{2E}$ are acquired by $\mathcal{M}$ and cancelled by four factors of $2E$ introduced for that purpose into the phase space.

We would like an exact result, containing the angular distribution, correct factors of $2\pi$ and so on. We can get it very easily because as $m/E \to 0$ spin 1 interactions couple through $\boldsymbol{\alpha}$, 1 which are diagonal between left and right handed states. In scattering $L \to L$, $R \to R$ while in annihilation or pair creation a left handed fermion annihilates with a right handed antifermion $L\bar{R} \to L\bar{R}$, $\bar{L}R$. This is what all the $\gamma$ algebra does for us.

In the centre of mass of an annihilating pair, the only non-zero component of angular momentum along the beam direction is contributed by the spins—aligned along this axis. The initial and final configurations are shown in fig.9.2

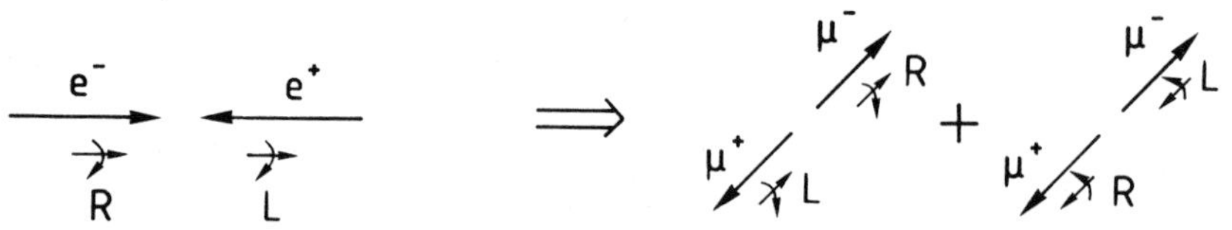


Fig.9.2

We add the amplitudes for $R\bar{L} \to R\bar{L}$ and square, then add the amplitudes for $R\bar{L} \to L\bar{R}$ and square. Adding the two squared amplitudes sums over all final states for $R\bar{L}$ in. The amplitudes are obtained by projecting the handedness states on to the $z$ axis where both spin components must point in the positive $z$ direction. The left handed particle state decomposes into

$$-\sin\frac{\theta}{2}\begin{pmatrix}1\\0\end{pmatrix} + \cos\frac{\theta}{2}\begin{pmatrix}0\\1\end{pmatrix} \tag{9.3}$$

relative to an axis making an angle $\theta$ with the momentum vector, and the right handed state is represented by

$$\cos\frac{\theta}{2}\begin{pmatrix}1\\0\end{pmatrix} + \sin\frac{\theta}{2}\begin{pmatrix}0\\1\end{pmatrix} \tag{9.4}$$

With two fermions, the amplitude for the final configuration $R\bar{L}$ is $2\cos^2\frac{\theta}{2}$ and for $L\bar{R}$ it is $2\sin^2\frac{\theta}{2}$. The factors of two are present because there are four $\gamma$ matrices; two are diagonal in spin and two off diagonal.

Thus summing over final states gives for the spinor part of the square of the matrix element for $R\bar{L}$ in

$$4\cos^4\frac{\theta}{2} + 4\sin^4\frac{\theta}{2} \tag{9.5}$$

There are two configurations of incoming spins which couple, out of 4 in total, so averaging over initial spin configurations yields

$$\overline{\Sigma|(\bar{u}'\gamma_\mu u')(\bar{u}\gamma_\mu u)|^2} = 1 + \cos^2\theta \tag{9.6}$$

for unpolarised beams. [Normalisation $u^+u = 1$—we were after the angular factors.]

The annihilation takes place from states with $J = 1$, $J_z = \pm 1$ with respect to the colliding beams, $C = -1$, $P = -1$ so you can see that if you produced a pair of spin zero particles the angular distribution would be $\sin^2\theta$.

We can introduce four factors of $\sqrt{2E}$ in $\mathcal{M}$ if we wish, but they will cancel with four factors of $2E$ in the Lorentz invariant phase space and flux factor. The point of doing so is that $(4E^2)^2\,(1+\cos^2\theta)$ can be written as a sum of scalar products of the incoming and outgoing 4-momenta, which is manifestly invariant.

The matrix element also contains the couplings and photon propagator, so that

$$|\mathcal{M}|^2 = \left(\frac{4\pi e^2}{q^2}\right)^2 (1+\cos^2\theta) \tag{9.7}$$

and with an incoming relative velocity of $2c$, the flux factor is $\frac{1}{2}$. Then

$$\frac{d\sigma}{d\Omega} = 2\pi \times \frac{1}{2} \times \left(\frac{4\pi e^2}{s}\right)^2 \times (1+\cos^2\theta) \times p^2 \frac{dp}{dE_{\mathrm{cm}}} \frac{1}{(2\pi)^3} \tag{9.8}$$

Now $E_{\mathrm{cm}} = 2p$, $dp/dE_{\mathrm{cm}} = \frac{1}{2}$, $p^2 = s/4$, so that

$$\frac{d\sigma}{d\Omega} = \frac{e^4}{4s}(1+\cos^2\theta) \tag{9.9}$$

and integrating over the full solid angle

$$\sigma = \frac{4\pi}{3}\frac{e^4}{s} \tag{9.10}$$

These two expressions (9.9) and (9.10) are among the most important in present day particle physics.

### 9.2 Fermion-fermion scattering

The crossed diagram, fig.9.3,

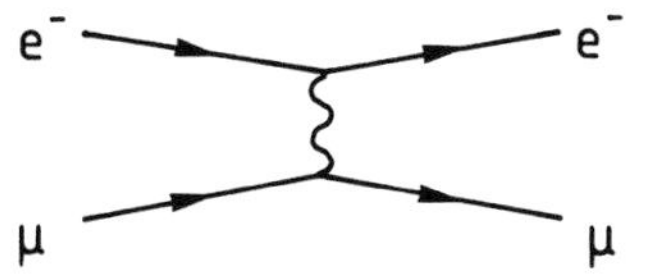


Fig.9.3

represents elastic scattering of two (non-identical) fermions by single photon exchange. The structure of the matrix element is identical to that in $e^+e^-$ annihilation—all that is necessary is to reverse a couple of 4-momenta to turn antiparticles into particles. But again we can get everything very simply in the centre of mass in the high energy limit. Handedness is conserved for any particle and there are two classes of initial spin configuration: same hands and opposite hands. The latter is marginally easier. For opposite hands the initial and final states are as shown in fig.9.4

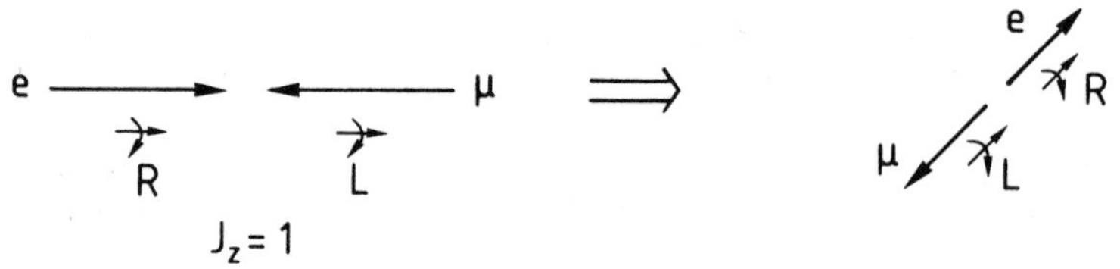


Fig.9.4

Project on the $J_z = 1$ final state and each particle contributes a factor of $\cos\frac{\theta}{2}$. The two $\gamma$ matrices diagonal in spin give a total amplitude for $RL \to RL$ of $2\cos^2\frac{\theta}{2}$.

For the cases where the two particles have the same handedness, $J_z = 0$ (fig.9.5)

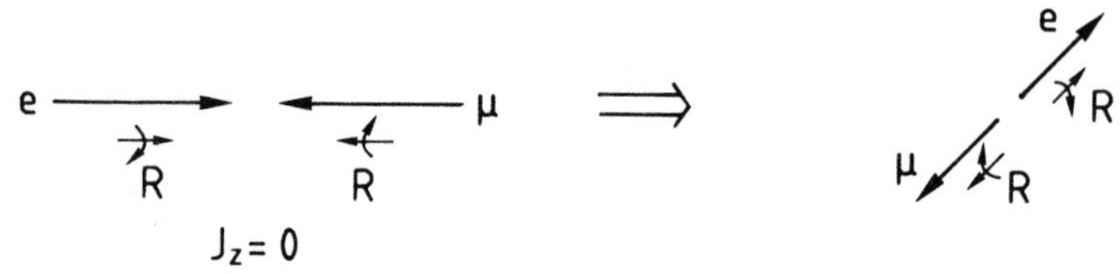


Fig.9.5

Obviously a term in the amplitude $2\cos^2\frac{\theta}{2}$ exists. In this case however, the $J_z = 0$ condition requires that the handed states are also projected upon the opposite combination of spin states relative to the $z$ axis: $\uparrow\downarrow + \downarrow\uparrow$. The latter contributes $2\sin^2\frac{\theta}{2}$, to be added in the amplitude to $2\cos^2\frac{\theta}{2}$ because both are contributed by the $RR$ final state. The vertex factor amplitude for $RR \to RR$ is isotropic and the value is 2. Note that (i) the two spin states are combined with the same symmetry as for the $RL$ case (ii) in the forward direction the scattering amplitudes for $RR$ and $RL$ become the same, as we would expect. Again the factors of 2 are because there are two $\gamma$ matrices diagonal in spin and two off diagonal. [$\boldsymbol{\alpha}$ has two components off diagonal between $u^+$ and $u$ and one component diagonal: the unit matrix is diagonal.] Summing over all final states and averaging over the 4 initial states

$$\begin{aligned}\overline{\Sigma|(\bar{u}'\gamma_\mu u')(\bar{u}\gamma_\mu u)|^2} &= 4(1+\cos^4\frac{\theta}{2}) \times 2 \div 4 \\ &= 2(1+\cos^4\frac{\theta}{2})\end{aligned} \tag{9.11}$$

Then in the centre of mass the cross section for $e\mu$ scattering by one photon exchange is given by

$$\frac{d\sigma}{d\Omega} = 2\pi \times \frac{1}{2} \times \left(\frac{4\pi e^2}{t}\right)^2 \times \left[2(1+\cos^4\frac{\theta}{2})\right] \times p^2\frac{dp}{dE_{\rm cm}}\frac{1}{(2\pi)^3} \tag{9.12}$$

($t$ is a popular label for $q^2$ in the momentum-like region). This result can be checked by taking the cleaner argument for $e^+e^-$ annihilation, expressing

$(2E)^4(1+\cos^2\theta)$ in terms of invariant products of 4-momentum vectors, and reversing the antiparticle 4-momenta. (The definition of $\theta$ is also changed by $\pi$.)

$$\cos^2\frac{\theta}{2} = \frac{1}{2}(1+\cos\theta) = 1-y \tag{9.13}$$

Here the angle $\theta$ is explicitly defined to be a centre of mass angle and therefore $\frac{1}{2}(1+\cos\theta)$ is an invariant. It has a special name, $1-y$, and is determined by $q^2$ and $s$.

With this definition,

$$d\cos\theta = (-)2dy \qquad d\Omega = 2\pi d\cos\theta$$

We also have the standard high energy relations

$$p^2 = s/4; \quad dp/dE_{\text{cm}} = \frac{1}{2}.$$

Then

$$\frac{d\sigma}{dy} = \frac{2\pi e^4}{(q^2)^2}\left[1+(1-y)^2\right]s \tag{9.14}$$

The first term in square brackets is contributed by scattering of fermions with the same handedness, the second term by opposite hands.

We have never studied $e\mu$ scattering at high centre of mass energies, for entirely practical reasons. However, replace the muon by a quark, which we take to be a spin 1/2 point particle, light, and satisfying the Dirac equation. The charge is $e_q$ in units of the electron charge $e$. (The quark target is of course a nucleon.)

Suppose a given species of quark carries a fraction $x$ of the nucleon momentum. The variable $s$ is the square of the centre of mass energy in the electron-quark system,

$$\begin{aligned} s &= (E_q+E_e)^2 - (\mathbf{p}_q+\mathbf{p}_e)^2 \\ &= (E_q+E_e)^2 - (p_q-p_e)^2 = 4E_eE_q = 4E_eE_Nx \end{aligned} \tag{9.15$a$}$$

neglecting masses. Denote the square of the centre of mass energy of the electron-nucleon system by

$$\begin{aligned} S &= (E_N+E_e)^2 - (\mathbf{p}_N+\mathbf{p}_e)^2 \\ &= (E_N+E_e)^2 - (p_N-p_e)^2 = 4E_NE_e \end{aligned} \tag{9.15$b$}$$

Comparison of (9.15a) and (9.15b) yields

$$s = xS \tag{9.16}$$

The differential cross section for scattering from effectively free quarks in the nucleon is given by

$$\frac{d\sigma}{dxdy} = 2\pi\left(\frac{e^2}{q^2}\right)^2[1+(1-y)^2]S\sum_q e_q^2 f_q(x)x \tag{9.17}$$

for the scattering of unpolarised electrons or muons (or their antiparticles) from an unpolarised target, under conditions of deep inelastic lepton-nucleon scattering. The data are in agreement with this expression—deep inelastic scattering of leptons from nucleons is elastic scattering of leptons from spin 1/2 quarks. The factor $[1+(1-y)^2]$ is the characteristic of spin 1/2 partons, the quarks.

### 9.3 Deep inelastic neutrino scattering

The same treatment applies to deep inelastic scattering of neutrinos. The fundamental processes are

$$\begin{array}{ll} \nu_\mu + d \to u + \mu^- & \bar{\nu}_\mu + u \to d + \mu^+ \\ [\nu_\mu + \bar{u} \to \bar{d} + \mu^- & \bar{\nu}_\mu + \bar{d} \to \bar{u} + \mu^+] \end{array}$$

The (charge changing) weak interactions only involve left handed particles and right handed antiparticles. The initial state $\nu q$ is therefore $LL \to LL$, same hands, and the vertex factors are isotropic in the centre of mass. The initial state $\bar{\nu}q$ is $\bar{R}L$, opposite hands, and the $\mu^+$ acquires an angular distribution $(1+\cos\theta)^2$ from the vertex factors, relative to the $\bar{\nu}$ direction. With the conventional definition of the Fermi coupling constant (as in nucleon $\beta$ decay) the electromagnetic propagator and coupling is replaced (9.18)

$$\frac{4\pi e^2}{q^2} \longrightarrow \frac{4G_F}{\sqrt{2}} \tag{9.18}$$

[The ugly choice (9.18) is largely historical in origin. The weak interactions were first studied in nuclear $\beta$ decay and a great quantity of data existed before the discovery of parity violation in 1957. For $q^2 \ll M_W^2$ we would now write the weak interaction matrix element in the form

$$\frac{4\pi g^2}{M_W^2}\left[\bar{\psi}\gamma_\mu\frac{(1-\gamma_5)}{2}\psi\right]\left[\bar{\psi}\gamma_\mu\frac{(1-\gamma_5)}{2}\psi\right]$$

where the factors $(1-\gamma_5)/2$ project out the left handed components of fermions,

$$\frac{(1-\gamma_5)}{2}\psi_L = \psi_L; \qquad \frac{(1-\gamma_5)}{2}\psi_R = 0$$

It would be convenient to make the substitution

$$\frac{4\pi e^2}{q^2} \to \frac{4\pi g^2}{M_W^2} = G$$

but the Fermi coupling constant $G_F$ is defined in a way determined by nuclear physics. Suppose that the decay rate for a super-allowed Fermi transition is measured, for example $^{14}O \to {}^{14}N^*$. The nuclear states are both $0^+$ so the interaction becomes

$$\frac{G}{4}[\bar{\psi}\gamma_\mu(1-\gamma_5)\psi][\bar{\Psi}\gamma_\mu\Psi]$$

The Fermi constant was originally defined assuming an interaction

$$G_F[\bar{\psi}\gamma_\mu\psi][\bar{\Psi}\gamma_\mu\Psi]$$

The origin of the factor of 4 in (9.18) is now obvious. For the same decay rate,

$$2\left(\frac{G}{4}\right)^2 = G_F^2$$

because the factor $1-\gamma_5$ contributes an additional factor of 2 in the rate, hence (9.18). The charged current interaction is therefore taken to be

$$\frac{4G_F}{\sqrt{2}}\left[\bar{\psi}\gamma_\mu\frac{(1-\gamma_5)}{2}\psi\right]\left[\bar{\psi}\gamma_\mu\frac{(1-\gamma_5)}{2}\psi\right]$$

and with this definition $G_F$ is determined from muon decay.]

With these changes, we obtain from (9.17)

$$\frac{d\sigma_{\nu d(\bar{\nu}\bar{d})}}{dxdy} = \frac{1}{\pi}G_F^2 Sxf_d(x) \quad ; \quad \int_0^1 dy = 1 \tag{9.19}$$

$$\frac{d\sigma_{\bar{\nu}u(\nu\bar{u})}}{dxdy} = \frac{1}{\pi}G_F^2(1-y)^2 Sxf_u(x) \quad ; \quad \int_0^1 (1-y)^2 dy = \frac{1}{3} \tag{9.20}$$

where $G_F \simeq 10^{-5}\,\text{GeV}^{-2}$ $[q^2 \ll M_W^2]$.

The additional selectivity of the parity violating neutrino interactions gives us tremendous leverage to establish the consistency of the whole picture and measure the structure functions $f_q(x)$.

## 9.4 Experimental aspects

In order to study $e^+e^-$ annihilation into hadrons, colliding $e^+e^-$ beams are necessary, and the collisions must take place in the core of a detector which records charged particle trajectories. The more information available on the neutral hadrons produced, the better. Electron beams are relatively easy to make. Positrons are made by pair creation, collected, compressed, cooled and finally accelerated—but the stored circulated beams have to remain on target for hours.

Deep inelastic scattering of electrons is comparatively simple to study. The cross section (9.17) is a function only of the initial and final electron momentum. Electron beams are easy to make, charged particle directions and momenta are easily measured with a magnetic spectrometer, and electrons are easy to identify. The virtue of positron deep inelastic scattering is to provide a check on the validity of the one photon exchange approximation, under which $e^+$ and $e^-$ cross sections are identical. Most deep inelastic electron scattering was done at SLAC, with electron beams of up to $\sim$ 30 GeV laboratory energy. Muon beams are produced from the decay of relatively well collimated pion beams, reaching energies of $\lesssim$ 300 GeV at the Fermilab and CERN SPS accelerators.

Neutrino beams present more of a problem. They consist mostly of $\nu_\mu$, $\bar{\nu}_\mu$ produced (like muons) in the decays $\pi^+ \to \mu^+\nu_\mu$; $\pi^- \to \mu^-\bar{\nu}_\mu$. The pion beams can be charge selected to pick $\nu_\mu$ or $\bar{\nu}_\mu$, and both collimated and momentum selected, but this cannot be pushed too far before intensity is lost. While the direction of the incoming neutrino is well known, in general the neutrino energy is poorly known. The direction and momentum of the outgoing muon is easy, and muons are identified by their ability to penetrate (at high energy) substantial amounts of matter. The neutrino energy can only be obtained by measuring both the muon energy and the total energy in the hadronic debris. The majority of neutrino experiments employed heavy liquid bubble chambers, or sampling calorimeters which give a measure of the energy in the hadrons. Such detectors also serve as the target for neutrinos, and contain (approximately) the same number of protons as neutrons—isoscalar targets. Useful neutrino beams have been produced with energies up to $\sim 100$ GeV.

Given the lepton energy $E$ in the laboratory, the scattered energy $E'$ and the laboratory scattering angle $\theta$, we can calculate the kinematic variables $q^2$, $x$, $y$ rather easily. Here is one way (we neglect the lepton masses.)

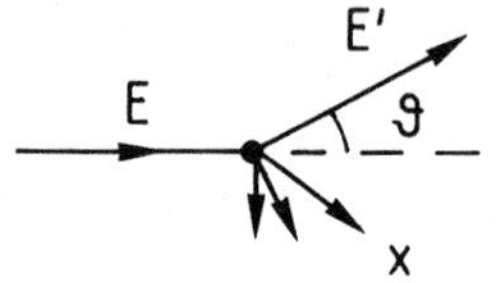


Fig.9.6

The quark-lepton centre of mass frame is reached by a Lorentz transformation in the beam direction such that $E_{\mathrm{cm}} = E'_{\mathrm{cm}}$, where $E_{\mathrm{cm}}$ is the energy of one particle in the cm frame.

$$E_{\mathrm{cm}} = E(1-\beta)\gamma \tag{9.21}$$
$$E'_{\mathrm{cm}} = E'(1-\beta\cos\theta)\gamma \tag{9.22}$$

Equating (9.21), (9.22) determines the boost parameter $\beta$

$$\beta = \frac{E-E'}{E-E'\cos\theta} \leq 1 \quad (E' \leq E) \tag{9.23}$$

The invariant $q^2$ is given by

$$\begin{aligned} q^2 = \Delta p^2 - \Delta E^2 &= (E-E'\cos\theta)^2 + (E'\sin\theta)^2 - (E-E')^2 \\ &= 2EE'(1-\cos\theta) = 4EE'\sin^2\frac{\theta}{2} \end{aligned} \tag{9.24}$$

The fractional momentum $x$ is given by

$$x = \frac{p_{q_{\mathrm{cm}}}}{p_{N_{\mathrm{cm}}}} \tag{9.25}$$

where

$$p_{q_{\mathrm{cm}}} = E_{\mathrm{cm}} = E(1-\beta)\gamma$$

and

$$p_{N_{\rm cm}} = M\beta\gamma$$

Thus

$$x = \frac{E}{M}\frac{1-\beta}{\beta} = \frac{EE'(1-\cos\theta)}{M(E-E')} = \frac{q^2}{2M\nu} \tag{9.26}$$

where $q^2$ is defined positive and $\nu = E - E'$.

Finally we may obtain $y$, defined by

$$1 - y = \frac{1}{2}(1 + \cos\theta_{\rm cm}).$$

Evaluate $q^2$ in the lepton-quark centre of mass. It is given by

$$q^2 = 2E^2_{\rm cm}(1 - \cos\theta_{\rm cm}) = \frac{s}{2}(1 - \cos\theta_{\rm cm}) \tag{9.27}$$

Then

$$\frac{1}{2}(1 - \cos\theta_{\rm cm}) = q^2/s = q^2/Sx \tag{9.28}$$

$$\frac{1}{2}(1 + \cos\theta_{\rm cm}) = 1 - q^2/Sx = 1 - \frac{2M\nu}{S} = 1 - \frac{\nu}{E} \tag{9.29}$$

since $S = 2EM$. Then

$$1 - y = 1 - \nu/E; \quad y = \frac{E - E'}{E}. \tag{9.30}$$

The deep inelastic cross section is a function of two variables. One popular pair is $\theta(lab)$ and $E'$, but you can use $q^2$, $E'$ or the parton model set $x$, $y$ ...

The quantity $\Sigma e_q^2 x f_q(x)$ in (9.17) is known in the trade as $\nu W_2(q^2, \nu)$. The fact that it is a function only of $x$ reveals the point-like nature and effective freedom of the partons struck by virtual photons (or $W^\pm$). The variation with $y$ at fixed $x$ reveals that these partons have spin $1/2$ and behave (strictly speaking, in the infinite momentum frame) as light Dirac fermions. The agreement between the structure functions $f(x)$ determined by deep inelastic electron and muon scattering and those determined by $\nu$ and $\bar{\nu}$ scattering pins down the whole picture firmly and finally establishes that the electro-weak interacting partons are indeed the quarks. The quarks are light, behaving approximately as massless fermions in the centre of mass frame. We shall see that quarks do not carry all the momentum of the nucleon.

## Problems

9.1 Consider $e^+e^-$ annihilation to a fermion-antifermion pair (any pair other than $e^+e^-$). In obtaining (9.5) a universal coupling ($e$) was used for all fermions and for $R\bar{L}$ and $\bar{R}L$. Suppose that there exists a process for which left-handed and right-handed couplings are different and it is necessary to specify separately

$$g_e^L, \quad g_e^R \quad ; \quad g_f^L, \quad g_f^R$$

Calculate the angular distribution of the final fermions in the centre of mass.

[Such a process is parity violating and it exists. Annihilation through a virtual $Z^\circ$ has these features and interference between the dominant virtual photon amplitudes and the virtual $Z^\circ$ amplitudes induce a forward-backward asymmetry in $e^+e^-$ annihilation.]

9.2 Adopt the relativistic normalisation and hence write the matrix elements for

$$e^+e^- \to \mu^+\mu^- \quad ; \quad e^+\mu^- \to e^+\mu^-$$

in a manifestly covariant form, in terms of scalar products of four-vectors. Show that the manifestly covariant expressions for the two processes are related by $s \leftrightarrow t$ crossing.

9.3 Calculate explicitly by matrix multiplication the results given in eqs.(9.6) and (9.11). You only have to multiply $2 \times 2$ matrices, so the algebra is trivial. Remember to employ common axes throughout and be very careful to identify the correct expressions for the coupled currents, or you will get nonsense. Remember that a negative energy IN state with momentum $\mathbf{p}$ is reinterpreted as a positive energy OUT state with momentum $-\mathbf{p}$ but the same helicity.

[This problem should lay to rest any residual doubts about factors of 2 and is generally enlightening as an explicit calculation of a sum usually done by trace techniques.]

9.4 Consider a Fermi transition in nuclear $\beta$ decay leading to $e^+\nu_e$. Use the helicity states of (8.31) to evaluate

$$\sum_{spins} |\bar{u}\gamma_4 u|^2$$

and show that

$$\sum_{spins} |\bar{u}\gamma_4(1-\gamma_5)u|^2$$

has twice the value. This calculation makes explicit the origin of the factor $\frac{1}{\sqrt{2}}$ in (9.18).

[As in problem 8.4, take the $z$ axis along the electron direction.]

9.5 While it is easy to determine neutrino direction, it is hard to determine neutrino energy. It can be done at the cost of intensity. The 'narrow band' neutrino beam at CERN was produced in the following way. A 400 GeV proton beam bombarded a primary target, and an approximately monochromatic pencil beam of secondaries was extracted, with a momentum of 200 GeV. This pencil beam was directed through a decay tunnel 305m long, followed by an absorbing region consisting of 184m of steel and 170m of rock. A detector was located 385m beyond the end of the decay tunnel. The neutrino beam resulted from the decays

$$\pi \to \mu + \nu \quad ; \quad K \to \mu + \nu$$

Find the relation between the laboratory energy of a neutrino and the angle at which it is emitted in the laboratory, with respect to the direction of the parent particle. Hence find the maximum and minimum energies of neutrinos reaching the detector within a radius of 1m from the beam axis.

If the detector can be idealised as a cylinder of liquid hydrogen of radius 1m normal to the beam and length 2m, estimate the number of neutrino interactions if $10^{10}$ pions enter the decay tunnel (corresponding to one burst).

# 10. PROBING THE NUCLEON.

## 10.1 Charged lepton probes

Deep inelastic scattering of charged leptons from both protons and neutrons (the neutrons being almost free in a deuterium target) is readily studied—the subject is almost 20 years old. The important point is that the scattering is correctly given as a function of the kinematic variables $x$ and $y$ by the quark parton model: it is the variation with $y$ at fixed $x$ which establishes the spin ½ Dirac nature of the partons which interact electromagnetically and weakly. There is a lot of interesting and important detail.

For protons

$$xF_p(x) = \Sigma e_q^2 x f_q(x) = x\left\{\frac{4}{9}(u^p(x)+\bar{u}^p(x)) + \frac{1}{9}(d^p(x)+\bar{d}^p(x)) + \frac{1}{9}(s^p(x)+\bar{s}^p(x))\right\} \tag{10.1}$$

and for neutrons

$$xF_n(x) = \Sigma e_q^2 x f_q(x) = x\left\{\frac{4}{9}(u^n(x)+\bar{u}^n(x)) + \frac{1}{9}(d^n(x)+\bar{d}^n(x)) + \frac{1}{9}(s^n(x)+\bar{s}^n(x))\right\} \tag{10.2}$$

neglecting any possibility of $c\bar{c}$ ... pairs within the nucleon. The $s$, $\bar{s}$ quarks only exist within the nucleon in pairs, as a result of quantum fluctuations—so-called sea quarks. Pairs of $u\bar{u}$, $d\bar{d}$ are present in the same way, but the proton valence quarks are $uud$; neutron valence quarks $ddu$.

Define

$$\begin{aligned} u^n(x) &= d^p(x) = d(x) \\ d^n(x) &= u^p(x) = u(x) \end{aligned} \tag{10.3}$$

Certainly

$$\begin{aligned} \bar{u}^p(x) &= \bar{d}^p(x) = \bar{u}^n(x) = \bar{d}^n(x) \\ \bar{s}^p(x) &= s^p(x) = \bar{s}^n(x) = s^n(x) \end{aligned} \tag{10.4}$$

Let us hope that $\bar{s}^p(x) \simeq \bar{u}^p(x)$, despite the greater strange quark mass, and label all functions in (10.4) $s(x)$ where $s$ now stands for sea, not strange.

The proton structure function is then

$$F_p(x) = \left\{\frac{4}{9}(u(x)+s(x)) + \frac{1}{9}(d(x)+s(x)) + \frac{1}{9}(2s(x))\right\} \tag{10.5}$$

and

$$F_n(x) = \left\{\frac{4}{9}(d(x)+s(x)) + \frac{1}{9}(u(x)+s(x)) + \frac{1}{9}(2s(x))\right\} \tag{10.6}$$

Then

$$F_p(x) - F_n(x) = \frac{1}{3}[u(x)-d(x)] = \frac{1}{3}[u_v(x)-d_v(x)] \tag{10.7}$$

where $v$ stands for valence quarks.

The form of the functions $xF_p(x)$ and $xF_n(x)$, as determined from deep inelastic electron scattering at SLAC, are shown in fig.10.1(a),(b). These functions are large at small values of the variable $x$ : $F(x) \approx 1/x$ as $x$ becomes small and soft partons proliferate.

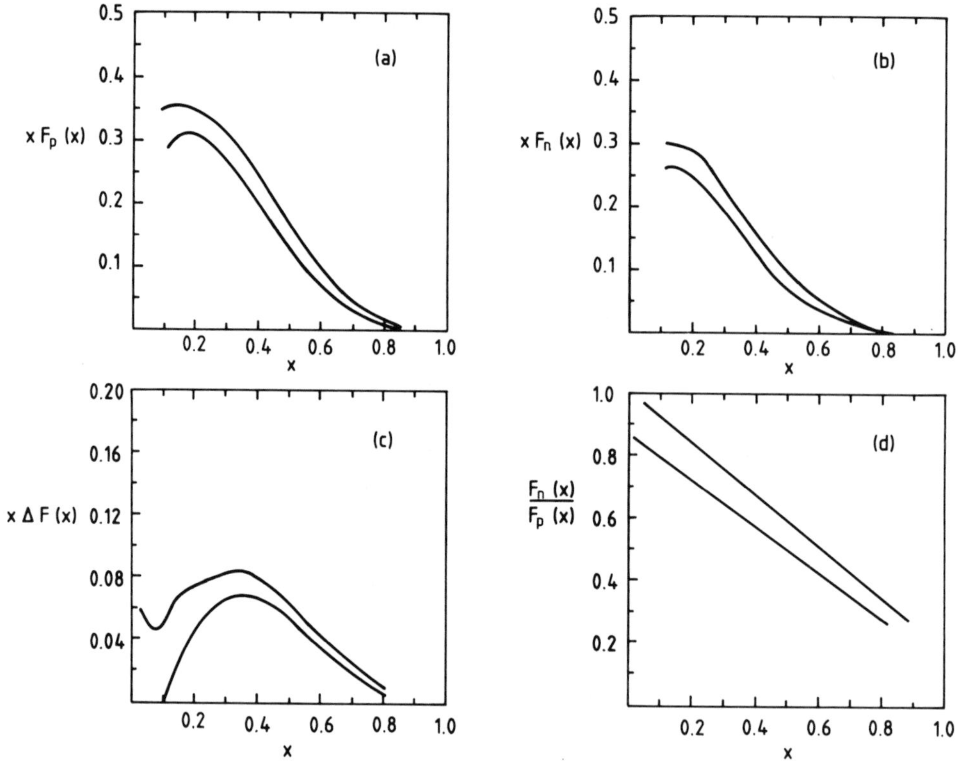


Fig.10.1

Proton and neutron electromagnetic structure functions, together with their difference and ratio. The two lines in each figure indicate the uncertainties due to errors and the scatter of data points. [From original data in A. Bodek *et al*, *Phys. Rev.* **D20** 1471 (1979); J.I. Friedman and H.W. Kendall, *Ann. Rev. Nucl. Sci.* **22** 203 (1972).]

The difference, eq.(10.7), is displayed in fig.10.1(c). It has the virtue that the sea quarks are eliminated, and to the extent that $u^p(x) \simeq 2d^p(x)$

$$F_p(x) - F_n(x) \simeq \frac{1}{3} q_v^1(x) \tag{10.8}$$

where $q_v^1(x)$ is the $x$ distribution of a single valence quark. You will note that the function

$$x\left[F_p(x) - F_n(x)\right] \tag{10.9}$$

shown in figure 10.1(c) is very different from $F_p(x)$, $F_n(x)$ separately, falling to approximately zero as $x \to 0$. The sea quarks have been eliminated in the difference. The function (10.9) peaks at $x \simeq \frac{1}{3}$ and it is tempting to interpret this as a manifestation of the three valence quarks in the nucleon. This is dangerous. If the nucleon consisted of three decoupled quarks, each carrying one third of the momentum of the nucleon, then both (10.7) and (10.9) would be a delta function at $x = \frac{1}{3}$, but the quarks are not decoupled and do not carry all the momentum of the nucleon, as we shall see. It is the case, however, that the integral

$$\int_0^1 [F_p(x) - F_n(x)]dx$$

should be equal to one third, and it is—within substantial errors arising largely from the uncertainty in (10.9) at low $x$ (see fig.10.1(c)).

The best measure of the sea quark population is extracted from neutrino deep inelastic scattering, section 10.2, but the difference (10.9) shows that sea quarks dominate at low $x$. Because the sea quark population must be the same in proton and neutron, we expect

$$F_n(x)/F_p(x) \to 1 \quad \text{as} \quad x \to 0$$

and it does, fig.10.1(d). Conversely, if only valence quarks are important at high $x$ we would expect, assuming $u^p(x) \simeq 2d^p(x)$ as $x \to 1$

$$\begin{aligned} F_p(x) &\to d(x) \\ F_n(x) &\to \frac{2}{3}d(x) \end{aligned}$$

This relation does not work. As $x \to 1$, $F_n(x)/F_p(x) \gtrsim \frac{1}{4}$, $u(x) \gg d(x)$ as $x \to 1$. This suggests a quark ($u$)-diquark ($ud$) structure for the proton. (Remember that the slope of $J$ *vs* $M^2$ for baryons also suggested a quark-diquark structure.)

Because $x$ represents the fraction of nucleon momentum carried by a quark, we would find

$$\sum_q \int_0^1 x f_q(x)dx = 1 \tag{10.10}$$

if quarks carried all the momentum of the nucleon.

$$\begin{aligned} \int_0^1 xF_p(x)dx &\approx \frac{4}{9}\epsilon_u + \frac{1}{9}\epsilon_d \\ \int_0^1 xF_n(x)dx &\approx \frac{4}{9}\epsilon_d + \frac{1}{9}\epsilon_u \end{aligned} \tag{10.11}$$

where $\epsilon_u$ is the fraction of proton momentum carried by $u(\bar{u})$ quarks and $\epsilon_d$ is the fraction of proton momentum carried by $d(\bar{d})$ quarks. The two integrals have been evaluated from data, and on solving for $\epsilon_u$, $\epsilon_d$

$$\epsilon_u \simeq 0.36 \qquad \epsilon_d \simeq 0.18$$

[Since there are two $u$ quarks and one $d$ in the proton, this result is reasonable.] Only half of the nucleon momentum is carried by the quarks. The rest must be carried by the gluons, or equivalently by the colour field—energy stored in the string. Energy stored in the fields will appear as momentum as well as energy on transforming to a frame in which the nucleon is moving with $v \sim c$. The concrete (but simple minded) string model gives a clear picture. Consider a single string, linking a quark and a diquark. In the rest frame of the nucleon we have

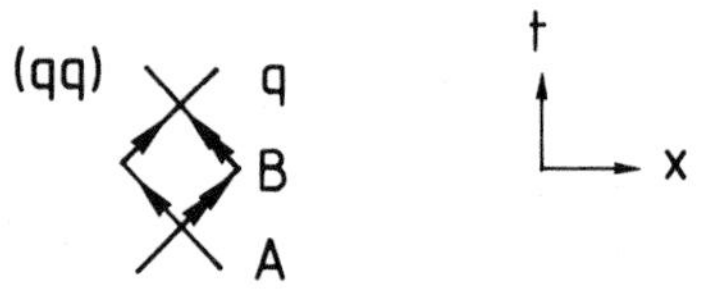


Fig.10.2

At $A$ the string is unstretched and all the energy is carried by the quarks. At $B$ all the energy is momentarily in the string, as the ends reverse their motions. On average half the energy is in the string, and the other half localised on the (massless) quark and diquark. Boost to a frame in which the string is moving very fast transverse to its length. The internal motions slow down $(1/\gamma)$ and the longitudinal momentum is equally shared between the quarks and the string, on average.

There is a curious fact concerning the structure functions observed using nuclei as targets. The deuteron is almost unbound, but probing heavy nuclei with muons ($A \sim 60$)

$$F_A(x)/A \neq F_D(x)/2$$

The valence quarks in a bound nucleus are degraded in momentum—the nuclear binding has resulted in more energy in the fields or in the sea. This is known as the EMC (European Muon Collaboration) effect.

### 10.2 Neutrino probes

The extraction of the fraction of nucleon momentum carried by quarks from electron or muon deep inelastic scattering is dependent on the assumed quark charges. If the assumed mean square charges are wrong the results are wrong. The structure functions can be independently measured using neutrino deep inelastic scattering, where the coupling to the quarks is known from $\beta$ decay (in which only one quark changes).

The (left handed) neutrino interacts only with $d$, $\bar{u}$

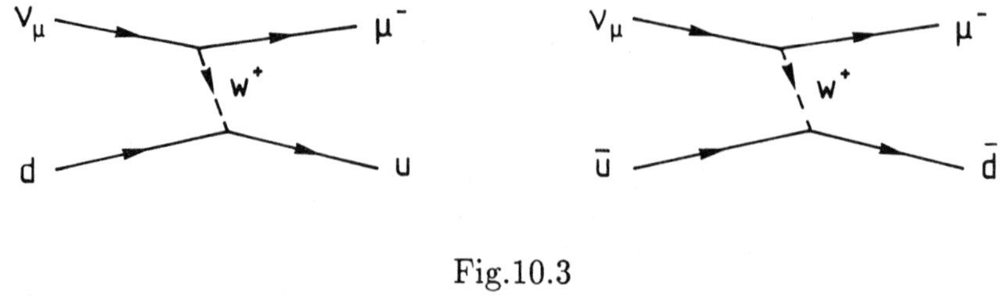


Fig.10.3

and the (right handed) antineutrino interacts only with $u$, $\bar{d}$. Since the $W^\pm$

couples to left handed particles and right handed antiparticles, the initial states

$$\begin{array}{llll} \nu q, \quad \bar{\nu}\bar{q} & \text{are} & LL, \quad \bar{R}\bar{R} & \text{- same handedness} \\ \bar{\nu}q, \quad \nu\bar{q} & \text{are} & \bar{R}L, \quad L\bar{R} & \text{- opposite handedness} \end{array}$$

The vertex factors for the same handedness pairs are isotropic, for the opposite handedness case a factor $(1-y)^2$ obtains, as we determined in Ch.9. In the framework of the quark parton model we can write down the cross sections for $\nu, \bar{\nu}$ on $p, n$ trivially:

$$\begin{aligned} \frac{d\sigma_{\nu p}}{dxdy} &= \frac{G_F^2}{\pi} Sx \left\{ d^p(x) + (1-y)^2 \bar{u}^p(x) \right\} \\ \frac{d\sigma_{\nu n}}{dxdy} &= \frac{G_F^2}{\pi} Sx \left\{ d^n(x) + (1-y)^2 \bar{u}^n(x) \right\} \\ \frac{d\sigma_{\bar{\nu} p}}{dxdy} &= \frac{G_F^2}{\pi} Sx \left\{ (1-y)^2 u^p(x) + \bar{d}^p(x) \right\} \\ \frac{d\sigma_{\bar{\nu} n}}{dxdy} &= \frac{G_F^2}{\pi} Sx \left\{ (1-y)^2 u^n(x) + \bar{d}^n(x) \right\} \end{aligned} \tag{10.12}$$

(neglecting $s, \bar{s}$ quarks)

For an (approximately) isoscalar target, add $\sigma_p$ and $\sigma_n$ and divide by 2 to obtain the cross section per nucleon:

$$\begin{aligned} \frac{d\sigma_{\nu N}}{dxdy} &= \frac{G_F^2}{2\pi} Sx \left\{ Q(x) + (1-y)^2 \bar{Q}(x) \right\} \\ \frac{d\sigma_{\bar{\nu} N}}{dxdy} &= \frac{G_F^2}{2\pi} Sx \left\{ (1-y)^2 Q(x) + \bar{Q}(x) \right\} \end{aligned} \tag{10.13}$$

where

$$\begin{aligned} Q(x) &= d^p(x) + d^n(x) = d^p(x) + u^p(x) = u^p(x) + u^n(x) \\ \bar{Q}(x) &= \bar{u}^p(x) + \bar{u}^n(x) = \bar{u}^p(x) + \bar{d}^p(x) \ldots \end{aligned} \tag{10.14}$$

For deep inelastic electron or muon scattering

$$\frac{d\sigma}{dxdy} = 2\pi \left( \frac{e^2}{q^2} \right)^2 \left[ 1 + (1-y)^2 \right] S \sum e_q^2 x f_q(x) \tag{10.15}$$

For an isoscalar target, the cross section per nucleon is obtained by adding the proton and neutron cross sections and dividing by 2. Replace $\sum e_q^2 f_q(x)$ with the average

$$\begin{aligned} &\frac{1}{2} \left\{ \frac{4}{9}(u^p(x) + u^n(x)) + \frac{4}{9}(\bar{u}^p(x) + \bar{u}^n(x)) \right. \\ &\quad \left. + \frac{1}{9}(d^p(x) + d^n(x)) + \frac{1}{9}(\bar{d}^p(x) + \bar{d}^n(x)) \right\} \\ &= \frac{5}{18} \left\{ Q(x) + \bar{Q}(x) \right\} \end{aligned} \tag{10.16}$$

Extract $Q(x)$, $\bar{Q}(x)$ from the $\nu$, $\bar{\nu}$ cross sections (and remember we know $G_F$ from nuclear $\beta$ decay where $d \to u$ or vice versa.) Compare their sum with the value extracted from $e$ or $\mu$ scattering. It works—the mean square quark charges are as in the fractionally charged model, just as for $e^+e^-$ annihilation (with colour).

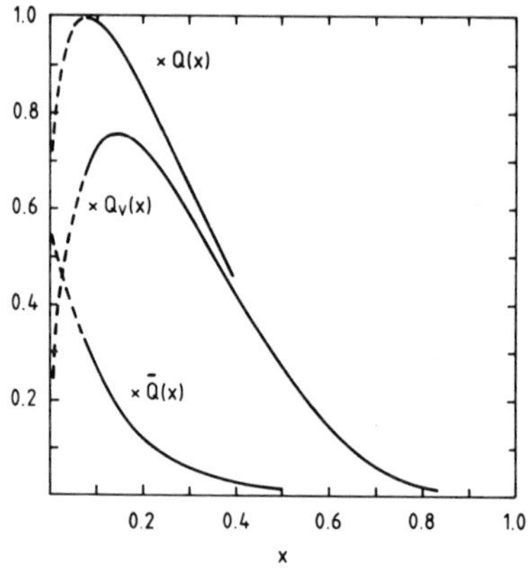


Fig.10.4

In fig.10.4 the functions $xQ(x)$, $x\bar{Q}(x)$ and the valence quark distribution $xQ_v(x)$ are shown. They have been constructed from the measured distributions known as $F_2^N(x)$, $xF_3^N(x)$ which, under the assumptions made to obtain (10.13, 10.16) may be identified with the sum and difference of the quark and antiquark distributions

$$\begin{aligned} F_2^N(x) &= x[Q(x) + \bar{Q}(x)] \\ xF_3^N(x) &= x[Q(x) - \bar{Q}(x)] = xQ_v(x) \end{aligned} \tag{10.17}$$

$F_2^N(x)$ was constructed using both deep inelastic muon scattering data and deep inelastic neutrino data; the agreement is excellent. The function $xF_3^N(x)$ was obtained from neutrino data alone. [See Particle Data Group, *Phys. Lett.* **170B** (1986) page 79.] The range of $q^2$ spanned was 10–30 (GeV)$^2$ and the valence quark structure function is noticeably softer than that extracted from SLAC data at lower $q^2$, fig.10.1(c).

The integral

$$\int_0^1 F_3^N(x)dx$$

should be equal to three, and it is (approximately). The integral

$$\int_0^1 F_2^N(x)dx$$

gives the fraction of nucleon momentum carried by quarks and antiquarks. It is again approximately one half. The neutrino data reveal the same fraction of nucleon momentum carried by gluons as is revealed by electron and muon scattering. The gluons couple to neither photons nor $W^{\pm}$.

Note that to the extent that $\int x\bar{Q}(x)dx$ can be neglected, integrating (10.13) over $x$ and $y$ gives

$$\sigma_{\nu N} = 3\sigma_{\bar{\nu} N}$$

which is the direct result of parity violation and the spin 1 couplings of $W^{\pm}$ to spin $1/2$. This is approximately correct, and the whole picture hangs together marvellously.

The structure functions are the relativistic equivalent of (non-relativistic) quark wave functions. It should be possible to calculate them from QCD ... one day. At present they are measured, but nonetheless very useful.

To calculate Drell-Yan cross sections for hadrons

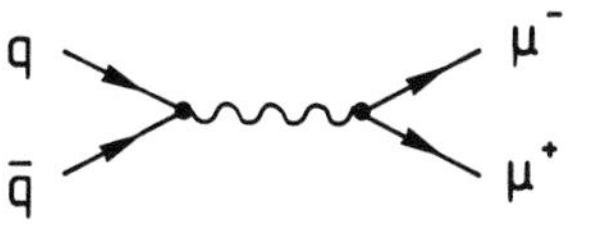


Fig.10.5

we need the one photon annihilation formula, plus the distribution of $q, \bar{q}$ momenta in the hadrons, the appropriate structure functions. It works very well for $\bar{p}N \to \mu^+\mu^- X$. Then measure $\pi N \to \mu^+\mu^- X$ and extract the pion structure functions. Similarly, to calculate the cross section for the process which led to observation of $W^{\pm}$

$$p + \bar{p} \to W^{\pm} + X$$

we need the weak quark-$W$ couplings and the quark and antiquark structure functions for $p$, $\bar{p}$—which we have as a result of more than 10 years work on lepton deep inelastic scattering.

In reality things are not that simple. The structure functions depend only on the variable $x$ in the most elementary models. As $q^2$ grows, the lepton probes smaller and smaller distance scales, and the structure of the fields surrounding the valence quarks becomes progressively more complicated. The contributions of the sea quarks and gluons increase logarithmically with $q^2$, at the expense of the momentum carried by the valence quarks, and the valence quark structure functions are softened. Bjorken scaling is violated. This evolution of structure functions with $q^2$ can be calculated in perturbative QCD through the (approximate) Altarelli-Parisi equations. [See for example Halzen and Martin *Quarks and Leptons* Wiley 1984.]

### 10.3 Weak neutral currents

Deep inelastic neutrino scattering probed the structure of the nucleon through the charged currents, but the same experiments probed the nature of the weak interaction in a unique way.

The weak interactions mediated by $W^{\pm}$ are called charged current reactions (in fact the weak current changes charge on interaction with $W^{\pm}$). In a bubble chamber picture such deep inelastic charged current events look like

Fig.10.6

Among the pictures in which such events were observed were some anomalies. Events containing no charged lepton were seen at $\sim 30\%$ of the rate for charged current events

Fig.10.7

and this was behind $\sim$ 500m of rock. If such events are due to neutrinos, then because of the absence of a charged lepton in the final state the process must be due to weak currents in which charge is not exchanged

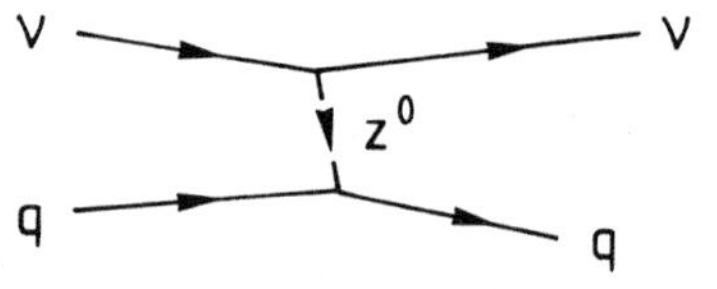


Fig.10.8

Only after years of painstaking work was it possible to be confident that all sources of possible background were eliminated, and that these are really weak neutral current processes. The clincher was the observation of neutrino-electron elastic scattering

$$\bar{\nu}_\mu + e^- \to \bar{\nu}_\mu + e^-$$

There is no diagram involving exchange of $W^\pm$ (and lepton number conservation) which can produce this. Neutral current ($Z^\circ$ exchange) processes can

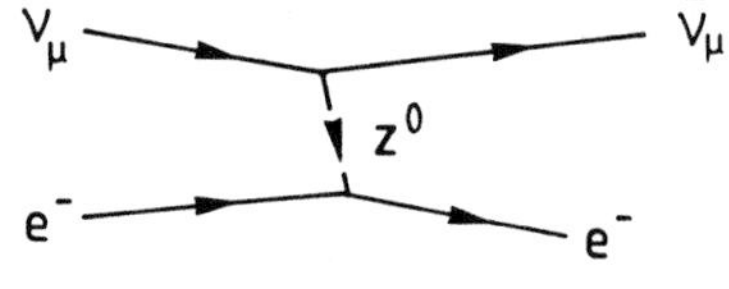


Fig.10.9

The experimental signature was electrons of energy $\gtrsim$ a few hundred MeV taking off $\sim$ dead forwards.

Today, more than a handful of events corresponding to

$$q\bar{q} \to Z^\circ \to \mu^+\mu^-, e^+e^-$$

have been observed at the CERN and Fermilab $p\bar{p}$ colliders, and production in $e^+e^-$ collisions is imminent. The detection of neutral current interactions some 15 years ago was a pleasant shock. Pleasant because the $Z^\circ$ diagrams could be used to cancel some divergences still left in the theory of weak interactions after the introduction of massive $W^\pm$, but a shock because strangeness changing neutral currents are certainly negligible. Strangeness changing charged currents exist of course—this is how the lightest strange particles decay:

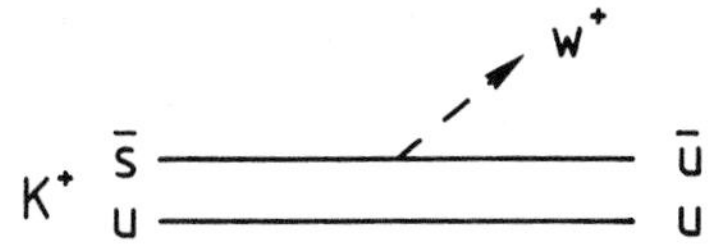


Fig.10.10

The decay $K^+ \rightarrow \mu^+ \nu_\mu$ has a branching ratio $\sim 0.64$ and an absolute rate $\sim 5\%$ of that expected if the $s\bar{u}$ coupling were equal to the $d\bar{u}$ coupling $G_F$. The diagram is

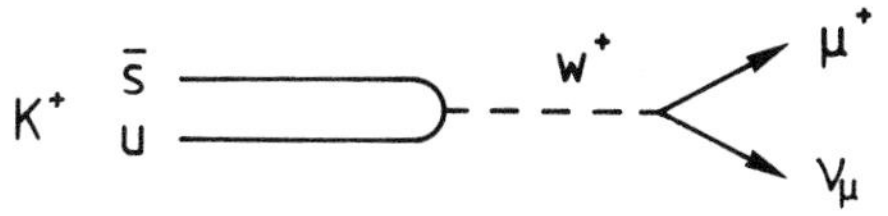


Fig.10.11

The analogous strangeness changing neutral current would be

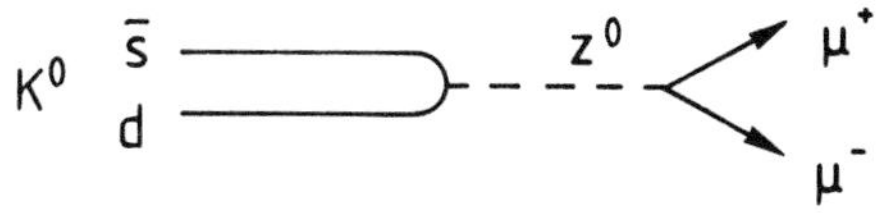


Fig.10.12

The branching ratio for $K^\circ \rightarrow \mu^+\mu^-$ is $\approx 10^{-8}$ and is a factor of at least $10^6$ too low for an $\bar{s}d$ coupling equal to $\bar{s}u$.

The rate at which $K^\circ - \bar{K}^\circ$ oscillations take place is also second order in the weak interactions and must proceed through diagrams such as fig.10.13(a) and not through 10.13(b)

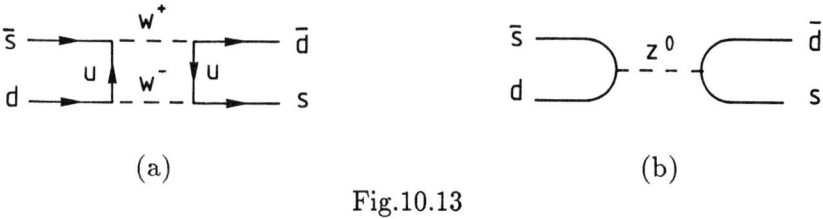

(a) (b)

Fig.10.13

There was of course no such experimental prohibition on neutral weak currents which do not change flavour, for in most circumstances processes such as

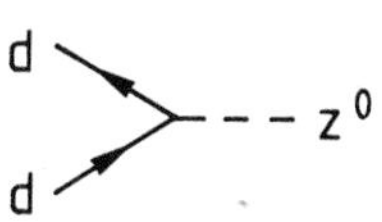


are swamped by

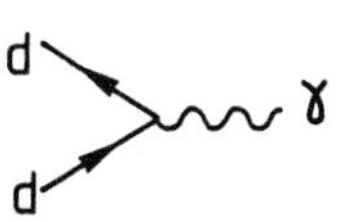


It is worth noting that the neutral currents offer a mechanism whereby a high flux of low energy neutrinos leaving the core of a supernova can exert significant pressure on the overlying layers.

The absence of strangeness changing neutral currents (and the suppression of such processes even at second order in the charged current processes) is (possibly) explained and (certainly) accomodated within the GIM mechanism and the existence of the second charge $+\frac{2}{3}$ quark, the charmed quark. One component of the GIM mechanism [Ch.15] is the existence of a flavour changing charged current interaction

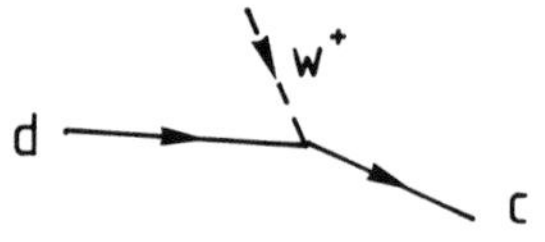


Fig.10.14

and in $\nu_\mu + N \to \mu^- + X$ charmed particles are produced singly, not in pairs, at a rate $\sim 10\%$. [There is also a mechanism $s \to c$, acting on sea quarks].

The first observation of a charmed particle was in fact made this way—although it was not known at the time that that was what it was.

## Problems

10.1 The functions $F_2^N(x)$ and $xF_3^N(x)$ have been parametrised as

$$F_2^N(x) = 3.9x^{0.55}(1-x)^{3.2} + 1.1(1-x)^8$$
$$xF_3^N(x) = 3.6x^{0.55}(1-x)^{3.2}$$

[see Particle Data Group *Phys. Lett.* **170B** (1986) page 79.] The data become uncertain below $x = 0.1$ and in determining the parameters no data below $x = 0.05$ were used. Assume that the above expressions are accurate over the full range of $x$ and evaluate the number of valence quarks in the nucleon and the fraction of nucleon momentum carried by quarks and antiquarks.

$xF_3^N$ is well determined above $x = 0.1$. Consider the extent to which $xF_3^N$ must differ from the above expression in order that the integral of $F_3^N$ be three, the number of valence quarks in the nucleon. To what extent would the calculated quark momentum be changed by such modifications?

10.2 To the extent that antiquarks can be neglected, (10.13) yields

$$\frac{\sigma_{\bar{\nu}N}}{\sigma_{\nu N}} = \frac{1}{3}$$

Estimate the correction to this relation as a result of including the effects of antiquarks present in the sea. [The measured number is approximately 0.38.]

10.3 Estimate the cross section for muon neutrino elastic scattering from electrons,

$$\bar{\nu}_\mu + e^- \to \bar{\nu}_\mu + e^-$$
$$\nu_\mu + e^- \to \nu_\mu + e^-$$

for neutrinos of energy 50 GeV (in the rest frame of the target electron). Compare the result with the cross section on nucleons. Consider how such events might be identified.

10.4 Estimate the cross sections for $W^+$ production in $p\bar{p}$ collisions at 540 and 1600 GeV in the centre of mass. The total $p\bar{p}$ cross section is $\sim$ 70mb. Calculate the total event rates and the $W^+$ event rates for a collider luminosity of $10^{29}\text{cm}^{-2}\text{s}^{-1}$.

10.5 Consider how the strange quark structure functions could be isolated from the quark-antiquark sea, in principle. Contemplate the practical difficulties.

# 11. SU(2) AND ISOSPIN.

## 11.1 Introduction

We now revert to the properties of hadrons at low energy, the regime in which quark binding is all important and the approximate representation of hadrons as clusters of free quarks is wholly inappropriate. The colour interactions are independent of flavour, and to the extent that the $u$ and $d$ quarks have identical mass, the physical states exhibit a $u \leftrightarrow d$ symmetry, isospin. The formal description is invariant under the operations of the group SU(2). Thus the hadrons exhibit a fairly good SU(2) of flavour symmetry–isospin–because the $u$ and $d$ quarks differ in mass only by ~MeV on a scale ~GeV. The strange quark mass is ~ 200 MeV and the physical states are not unchanged by operations such as $s \leftrightarrow d$; there is an SU(3) of flavour, but it is badly broken because of the strange quark mass. We may forget about the symmetries SU(4), SU(5), SU(6) of flavour entirely. They are of some use in classifying hadrons containing heavy quarks, but devoid of useful dynamical content.

Charge independence of nuclear forces can be expressed as conservation of isospin, but the formal development contains much convention which it is hard to separate from the essential physical content. In this chapter we examine the physics underlying an SU(2) symmetry—which could be isospin among the light hadrons, a weak isospin (remember that $W^{\pm}$ exchange charge and flavour) or a (non-existent) SU(2) of colour. The development of an SU(3) of colour to describe the interactions among quarks is then straightforward, and is the subject of Ch.12.

## 11.2 Exchange interactions again

Start with a basic doublet of particles of the same mass, distinguished by a two valued quantum number—labels $\alpha$, $\beta$ for generality. Require the existence of alphabetic exchange forces within the Hamiltonian, such that

$$\begin{aligned} H|\alpha\beta> &= E|\alpha\beta> + V|\beta\alpha> \\ H|\beta\alpha> &= E|\beta\alpha> + V|\alpha\beta> \end{aligned} \tag{11.1}$$

where the position in brackets in (11.1) denotes a position or momentum coordinate.

The eigenstates are trivial to obtain:

$$\begin{aligned} H|\alpha\beta + \beta\alpha> &= (E+V)|\alpha\beta + \beta\alpha> \\ H|\alpha\beta - \beta\alpha> &= (E-V)|\alpha\beta - \beta\alpha> \end{aligned} \tag{11.2}$$

An SU(2) has the property of charge independence for the elements of such doublets, just like the charge independence of nuclear forces ($\alpha, \beta \to p, n$) where the forces acting in any $p, n$ state labelled by spin-space coordinates are identical to those acting in $p, p$ and $n, n$ states with the same configuration (if permitted by the Pauli principle). Then one of the two alphabetic states $(\alpha\beta + \beta\alpha)$, $(\alpha\beta - \beta\alpha)$ is to have the same energy as $(\alpha\alpha)$, $(\beta\beta)$. It is both desirable and natural to choose $(\alpha\beta + \beta\alpha)$ degenerate with $(\alpha\alpha)$, $(\beta\beta)$. These three degenerate states then have the same alphabetic symmetry and we may treat the two basic elements $\alpha$, $\beta$ as two states of the same particle, with an extended Pauli principle (for fermions). We are not forced to make this elegant choice, but it would be perverse to choose otherwise.

Draw the operation of the exchange potential $V$ in the usual way:

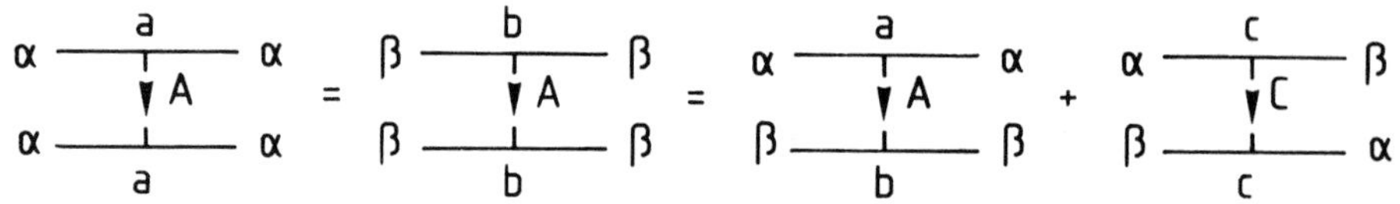


Fig.11.1

We require

$$a^2 = b^2 = ab + c^2 \tag{11.3}$$

when

$$\begin{aligned} &< \alpha\alpha|V|\alpha\alpha >= a^2 v \qquad\qquad < \beta\beta|V|\beta\beta >= b^2 v \\ &\frac{1}{2} < \alpha\beta + \beta\alpha|V|\alpha\beta + \beta\alpha > = (ab + c^2)v \\ &\frac{1}{2} < \alpha\beta - \beta\alpha|V|\alpha\beta - \beta\alpha > = (ab - c^2)v \end{aligned} \tag{11.4}$$

(The diagrams were set up with this choice — $(\alpha\beta + \beta\alpha)$ degenerate with $(\alpha\alpha)$ — in mind.) [If fig.11.1 is regarded as a set of Feynman diagrams, then it only represents scattering amplitudes in first order perturbation theory. It is nonetheless a convenient representation of the necessary relations between potentials, (11.3) and (11.4).]

The solutions are

| | | |
|---|---|---|
| $a = b,$ | $c = 0$ | alphabetic singlet exchange |
| $a = -b,$ | $c = \pm\sqrt{2}b$ | alphabetic triplet exchange |

Even choosing the three degenerate states to be all symmetric has not fixed all the phases and it is **conventional** to choose, with the definitions of fig.11.1

$$a = -\frac{1}{\sqrt{3}}, \qquad b = +\frac{1}{\sqrt{3}}, \qquad c = +\sqrt{\frac{2}{3}} \tag{11.5}$$

all multiplied by a common coupling constant. This choice is made because with the basic doublet $\binom{\alpha}{\beta}$ these numbers are just Clebsch-Gordan coefficients for $\frac{1}{2} \to \frac{1}{2} \otimes 1$ (angular momentum coupling coefficients) in the Condon and Shortley convention. In this convention a factor $-\sqrt{\frac{2}{3}}$ is associated with the $\beta \leftrightarrow \alpha\bar{C}$ vertex.

Notice that if the exchanged object $A$ is allotted alphabetic composition

$$A = \frac{1}{\sqrt{2}}(\beta\bar{\beta} - \alpha\bar{\alpha}) \tag{11.6}$$

and

$$C = \alpha\bar{\beta} \tag{11.7}$$

then the couplings are just proportional to the alphabetic projections at the appropriate vertices, while if $c = 0$ and $a = b$ the exchanged object has composition $\frac{1}{\sqrt{2}}(\alpha\bar{\alpha} + \beta\bar{\beta})$. This one is blind to $(\alpha, \beta)$ and is an alphabetic singlet. The structures $A$, $C$ and $\bar{C}$ are members of an alphabetic triplet. Then the alphabetic triplet potential gives the three $(\alpha, \beta)$ triplet states $+\frac{1}{3}V_T$ while the $(\alpha, \beta)$ singlet acquires $-V_T$. Singlet exchange of course leaves all four $(\alpha, \beta)$ states degenerate.

If we repeat the exercise for antiparticles labelled $\bar{\alpha}$, $\bar{\beta}$ then of course $\bar{a}^2 = \bar{b}^2 = \bar{a}\bar{b} + \bar{c}^2$; $\bar{a} = -\bar{b}$, $\bar{c} = \pm\sqrt{2}\bar{b}$.

The energies acquired by triplet exchange among particle-antiparticle states are more interesting. For $(\alpha, \bar{\beta})$, $(\beta\bar{\alpha})$ ONLY $A$ exchange can take place (fig.11.2)

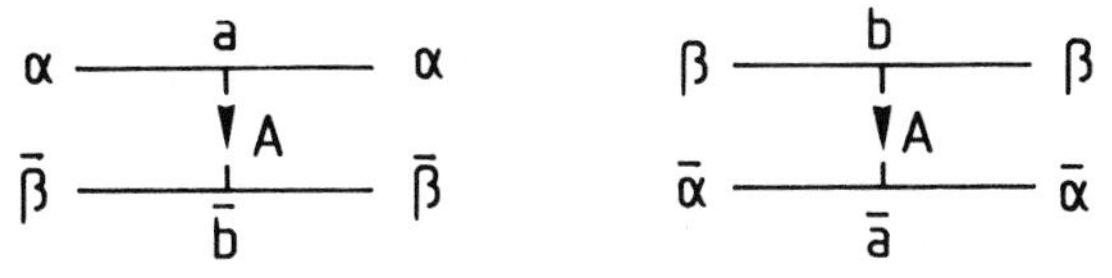


Fig.11.2

such that

$$\hat{V}_T|\alpha\bar{\beta}> \rightarrow a\bar{b}|\alpha\bar{\beta}> \qquad \text{and} \qquad \hat{V}_T|\beta\bar{\alpha}> \rightarrow b\bar{a}|\beta\bar{\alpha}> \tag{11.8}$$

Since $a\bar{b} = b\bar{a} = -b\bar{b}$, these two states are degenerate. It looks like they are two states of a triplet. The other two input states for particle-antiparticle are $\alpha\bar{\alpha}$, $\beta\bar{\beta}$, and here $C$ exchange is active in addition to $A$ exchange (fig.11.3)

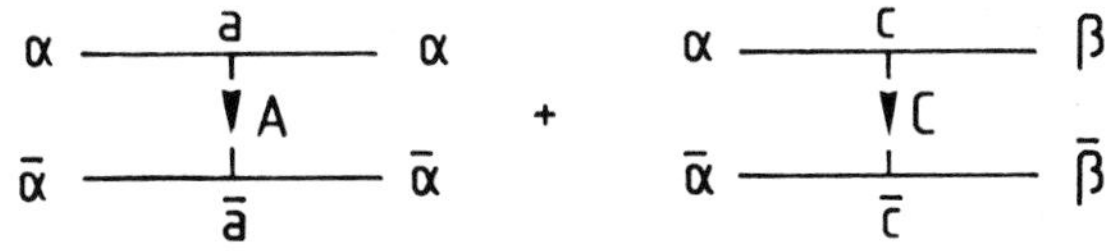


Fig.11.3

Thus

$$\begin{aligned} \hat{V}_T|\alpha\bar{\alpha}> &\rightarrow a\bar{a}|\alpha\bar{\alpha}> + c\bar{c}|\beta\bar{\beta}> \\ \hat{V}_T|\beta\bar{\beta}> &\rightarrow b\bar{b}|\beta\bar{\beta}> + c\bar{c}|\alpha\bar{\alpha}> \end{aligned} \tag{11.9}$$

Since $a = -b$ and $\bar{a} = -\bar{b}$, $a\bar{a} = b\bar{b}$ and then

$$\begin{aligned} \hat{V}_T|\alpha\bar{\alpha} + \beta\bar{\beta}> &\rightarrow (b\bar{b} + c\bar{c})|\alpha\bar{\alpha} + \beta\bar{\beta}> \\ \hat{V}_T|\beta\bar{\beta} - \alpha\bar{\alpha}> &\rightarrow (b\bar{b} - c\bar{c})|\beta\bar{\beta} - \alpha\bar{\alpha}> \end{aligned} \tag{11.10}$$

If we choose $c\bar{c} = +2b\bar{b}$, then

$$(\alpha\bar{\beta}), \frac{1}{\sqrt{2}}(\beta\bar{\beta} - \alpha\bar{\alpha}), (\beta\bar{\alpha})$$

are degenerate, and for each of these three states the expectation value of the triplet potential is $-b\bar{b}(\bar{v})$. The singlet state

$$\frac{1}{\sqrt{2}}(\alpha\bar{\alpha}+\beta\bar{\beta})$$

is split off, getting $+3b\bar{b}(\bar{v})$. With this choice, $c\bar{c}=+2b\bar{b}$, the (bound) states have the same alphabetic symmetry as the obvious allocation of quantum numbers to the exchanged objects in the $(\alpha,\beta)$ interaction, eqs.(11.6, 7). The state $\frac{1}{\sqrt{2}}(\alpha\bar{\alpha}+\beta\bar{\beta})$ is an SU(2) singlet (of, for example, isospin). The two mixed states would be reversed if the opposite arbitrary choice, $c\bar{c}=-2b\bar{b}$, were made. This is a possible choice but it is simpler, seems more natural, and is **conventional** that $\alpha\bar{\alpha}+\beta\bar{\beta}$ is the singlet, carrying the alphabetic structure of the vacuum.

### 11.3 Charge conjugation and SU(2)

The doublet

$$\begin{pmatrix}\alpha\\ \beta\end{pmatrix}\begin{matrix}+\frac{1}{2}\\ -\frac{1}{2}\end{matrix}$$

is converted by charge conjugation into the doublet $\begin{pmatrix}\bar{\alpha}\\ \bar{\beta}\end{pmatrix}$, but this is not taken as a basic doublet of alphabetic SU(2). Rather, one assigns the eigenvalues $\pm\frac{1}{2}$:

$$\begin{pmatrix}\bar{\beta}\\ -\bar{\alpha}\end{pmatrix}\begin{matrix}+\frac{1}{2}\\ -\frac{1}{2}\end{matrix}.$$

The reason is that then the same table of Clebsch-Gordan coefficients applies for manipulating particle-particle, antiparticle-antiparticle and particle-antiparticle interactions. One example is entirely obvious:

| particle-particle singlet | particle-antiparticle singlet |
|---|---|
| $\alpha\beta-\beta\alpha$ | $\bar{\beta}\beta+\bar{\alpha}\alpha$ |
| $\uparrow\downarrow-\downarrow\uparrow$ | $\uparrow\downarrow-\downarrow\uparrow$ |

where the arrows denote eigenvalues $\pm\frac{1}{2}$. Thus for isospin $T=\frac{1}{2}$ states we identify the eigenstates of $T_z$ according to (11.11):

$$\begin{pmatrix}p\\ n\end{pmatrix},\begin{pmatrix}\bar{n}\\ -\bar{p}\end{pmatrix}\ ;\ \begin{pmatrix}u\\ d\end{pmatrix},\begin{pmatrix}\bar{d}\\ -\bar{u}\end{pmatrix}\quad\begin{matrix}T_z=+\frac{1}{2}\\ T_z=-\frac{1}{2}\end{matrix}\qquad(11.11)$$

and this is no more than a convenient convention. [Note that $(-\beta\bar{\alpha})$ is to be identified with the (alphabetic) $T_z=-1$ particle-antiparticle state. Thus if $\bar{C}$ is exchanged rather than $C$, projection of the alphabetic content yields the factor $-\sqrt{\frac{2}{3}}$ associated with the $\beta\leftrightarrow\alpha\bar{C}$ vertex.]

Here is an example where apparently it is more than just a convention. The vector mesons $\rho^\circ$, $\omega^\circ$, $\phi$ are identified with $s$ wave spin triplets and have

$J^{PC} = 1^{--}$, the quantum numbers of the photon. They decay to $e^+e^-$, $\mu^+\mu^-$ via the process

Fig.11.4

(with branching ratios $\sim 10^{-4}$) and may be made in $e^+e^-$ annihilation by its inverse. Virtual vector mesons may also be made from a real photon and elastically scattered into the real state. The $\rho^\circ$ is a member of an isospin triplet (of $u\bar{d}\ldots$), $\omega^\circ$ is a singlet and $\phi \sim s\bar{s}$ is a singlet. The $\omega^\circ$ and $\rho^\circ$ are almost degenerate; the strange quark content of $\omega^\circ$ can be neglected. The photon has quark couplings

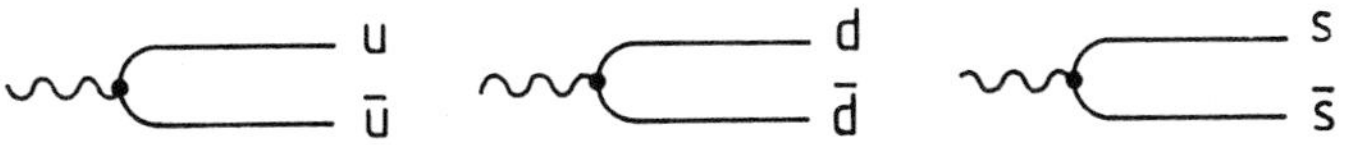


$$+\frac{2}{3}e \qquad\qquad -\frac{1}{3}e \qquad\qquad -\frac{1}{3}e$$

The isospin composition of the vector mesons is

$$\rho^\circ = \frac{1}{\sqrt{2}}(d\bar{d} - u\bar{u}) \qquad \omega^\circ = \frac{1}{\sqrt{2}}(u\bar{u} + d\bar{d}) \qquad \phi = s\bar{s}$$

Then

$$u\bar{u} = \frac{1}{\sqrt{2}}(\omega^\circ - \rho^\circ) \quad d\bar{d} = \frac{1}{\sqrt{2}}(\rho^\circ + \omega^\circ)$$

and

$$\gamma \longrightarrow \frac{2}{3}e\left\{\frac{\omega^\circ - \rho^\circ}{\sqrt{2}}\right\} - \frac{1}{3}e\left\{\frac{\rho^\circ + \omega^\circ}{\sqrt{2}}\right\} - \frac{1}{3}e\phi$$

$$= -\frac{e}{\sqrt{2}}\rho^\circ + \frac{e}{3\sqrt{2}}\omega^\circ - \frac{e}{3}\phi$$

To the extent that the quark wave functions are common, an assumption which should be good for $\rho^\circ$ and $\omega^\circ$, the squares of the photon-vector meson amplitudes should be in the ratio

$$|<\rho^\circ|\gamma>|^2 : |<\phi|\gamma>|^2 : |<\omega^\circ|\gamma>|^2 = 9:2:1$$

and of course they are. But at first sight had we chosen $u\bar{u} + d\bar{d}$ as the isospin triplet and $d\bar{d} - u\bar{u}$ as the singlet, the triplet ($\rho$) would then have a partial width $\frac{1}{9}$ of the singlet ($\omega$). This is not in fact the case. Had we made that choice, the vacuum would be represented not by the conventional choice $u\bar{u} + d\bar{d}$ but rather by the new singlet choice $d\bar{d} - u\bar{u}$. Then

$$\gamma|vacuum> \rightarrow \frac{2}{3}e|-u\bar{u}> - \frac{1}{3}e|d\bar{d}>$$

and with

$$\rho^\circ = \frac{1}{\sqrt{2}}(u\bar{u} + d\bar{d}), \quad \omega^\circ = \frac{1}{\sqrt{2}}(d\bar{d} - u\bar{u})$$

$$|<\rho^\circ|\gamma>|^2 : |<\omega^\circ|\gamma>|^2 = 9:1$$

Of course, applying this choice of isospin states consistently the state $u\bar{u} + d\bar{d}$ is degenerate with $u\bar{d}$, $\bar{u}d$, has the decay amplitudes characteristic of isospin 1, and so on. But we never make this inconvenient choice. The singlet (vacuum) structure is conventionally $u\bar{u} + d\bar{d}$ $(+s\bar{s} + \ldots)$.

### 11.4 Isospin selection rules

The $\rho^\circ$ has mass 770 MeV, width 150 MeV, and decays into $\pi^+\pi^-$. The $\omega^\circ$ has mass 783 MeV (almost degenerate with $\rho^\circ$), width 10 MeV and decays predominantly into $\pi^+\pi^\circ\pi^-$. Both are strong decays conserving isospin and $q\bar{q}$ states are held together by colour fields which are flavour independent. Decay takes place by creation of a $q\bar{q}$ pair and for isospin conservation the pair creation amplitude is $u\bar{u} + d\bar{d}$. [The $u$ and $d$ masses are only a few MeV different and so do not significantly affect the relative $u\bar{u}$, $d\bar{d}$ pair creation amplitudes.] Then for $\rho^\circ$ and $\omega^\circ$ decay we have pieces which may be represented by fig.11.5

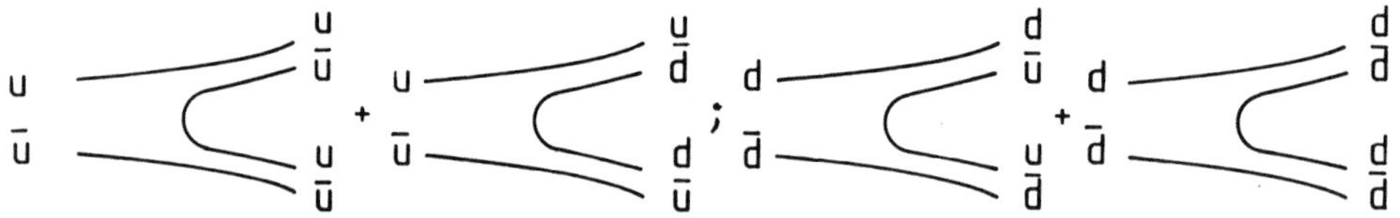

Fig.11.5

Denoting isospin states by $(T, T_z)$ we have

$$\frac{1}{\sqrt{2}}(u\bar{u} + d\bar{d}) = (0,0) \quad \frac{1}{\sqrt{2}}(d\bar{d} - u\bar{u}) = (1,0) \quad u\bar{d} = (1,+1) \quad \bar{u}d = -(1,-1) \tag{11.12}$$

Then

$$u\bar{u} = \frac{1}{\sqrt{2}}[(0,0) - (1,0)] \quad d\bar{d} = \frac{1}{\sqrt{2}}[(0,0) + (1,0)]$$

and since

$$u\bar{u} \to (u\bar{u})(u\bar{u}) + (u\bar{d})(d\bar{u})$$

we have

$$u\bar{u} \to \frac{1}{2}\{(0,0)(0,0) + (1,0)(1,0) - (0,0)(1,0) - (1,0)(0,0)\} - (1,+1)(1,-1) \tag{11.13}$$

and similarly

$$d\bar{d} \to \frac{1}{2}\{(0,0)(0,0) + (1,0)(1,0) + (0,0)(1,0) + (1,0)(0,0)\} - (1,-1)(1,+1) \tag{11.14}$$

Then

$$u\bar{u} + d\bar{d} = \sqrt{2}(0,0) \to (0,0)(0,0) - (1,+1)(1,-1) + (1,0)(1,0) - (1,-1)(1,+1) \quad (11.15)$$

$$d\bar{d} - u\bar{u} = \sqrt{2}(1,0) \to (0,0)(1,0) + (1,0)(0,0) + (1,+1)(1,-1) - (1,-1)(1,+1) \quad (11.16)$$

This gives us the isospin rules for

$$(0,0) \to 0 \otimes 0, 1 \otimes 1$$
$$(1,0) \to 1 \otimes 0, 1 \otimes 1$$

which can also be read from a table of Clebsch-Gordan coefficients. As expected, isospin zero does not decay into $1\otimes 0$ and isospin 1 does not decay into $0\otimes 0$—the amplitudes cancel between the $u\bar{u}$ and $d\bar{d}$ input pieces. You will also notice that $(1,0)$ does not decay into $(1,0)(1,0)$—the amplitudes cancelled. Thus isospin conservation forbids $\rho^\circ \to \pi^\circ\pi^\circ$. This decay mode is independently forbidden because two pions from decay of a $1^-$ meson must be in a $p$ wave and $\pi^\circ\pi^\circ$ is a pair of identical bosons, but the isospin rule is independent of this special case. There is however more to it.

### 11.5 Charge conjugation and $G$ parity

On the grounds of spin, parity and isospin conservation there is a reason why $\omega^\circ$ should not decay into two pions, for $\pi^\circ\pi^\circ$ is forbidden because the two $\pi^\circ$ are identical bosons, and the $T=0$ $\pi^+\pi^-$ amplitudes are equal in magnitude to the $\pi^\circ\pi^\circ$ amplitude. [$\omega^\circ$ does decay into $\pi^+\pi^-$ (1.4%) but only because of isospin breaking.] In $\rho^\circ$ decay, isospin gives a zero amplitude for $\pi^\circ\pi^\circ$ but not for $\pi^+\pi^-$. We have here an instance of a more general rule. The quark flavour composition determines isospin, and the spin, parity and charge conjugation number (of neutral mesons) are determined by the spin-space quark structure. The strong (colour) interactions are invariant under the charge conjugation operation ($C$) of turning particles into antiparticles. Together with isospin, this restricts the decay possibilities not only for eigenstates of $C$ ($\pi^\circ, \rho^\circ, \omega^\circ \ldots$) but also for charged meson states.

Without pushing the argument too far, we can see how the effect of $C$ invariance can be taken into account. Denote a $C=+1$ $u\bar{u}$ state as $u\bar{u}+\bar{u}u$ and a $C=-1$ state as $u\bar{u}-\bar{u}u$. The neutral pion is a (spin) singlet state with $C=+1$ and isospin $(1,0)$ so

$$\pi^\circ = \frac{1}{2}\left\{(d\bar{d} - u\bar{u}) + (\bar{d}d - \bar{u}u)\right\}$$

The $\pi^+$ has the same spin-space quark-antiquark configuration, which in $\pi^\circ$ is responsible for the value $C=+1$, so we represent the charged pions as

$$\pi^+ = \frac{1}{\sqrt{2}}(u\bar{d} + \bar{d}u) \qquad \pi^- = \frac{1}{\sqrt{2}}(d(-\bar{u}) + (-\bar{u})d)$$

Then $C\pi^+ = \frac{1}{\sqrt{2}}(\bar{u}d + d\bar{u}) = -\pi^-$ with the conventional definition of isospin states. We can now work out the implications of conservation of isospin AND $C$ by creating from the colour field the $C = +1$, $T = 0$ vacuum composition

$$u\bar{u} + \bar{u}u + d\bar{d} + \bar{d}d$$

and accepting bound hadron states:

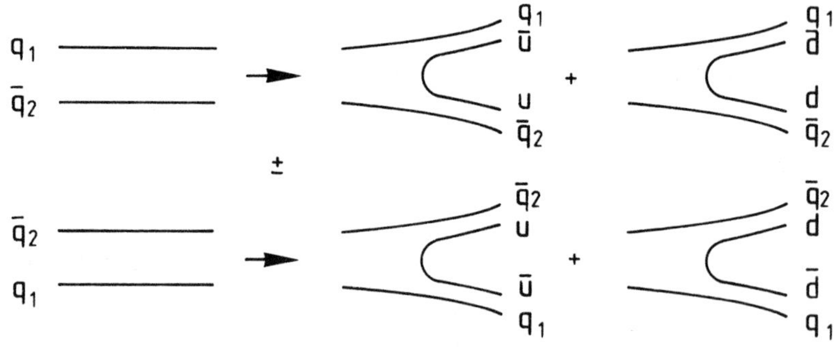


Fig.11.6

The initial state is constructed from the two input states above with a + or − sign according to $C$ for the neutral meson, and $q_1, \bar{q}_2$ chosen for $(T, T_z)$. The final states are then decomposed into eigenstates of $(T, T_z)$ and appropriate transformation properties under $C$. The rules obtained are just those of isospin (of course), plus one more. The effect of $C$ conservation for the strong interactions of isospin 0 and 1 mesons is summarised by defining a new multiplicative quantum number $G$ (called $G$ parity), where

$$G = C_n(-1)^T \tag{11.17}$$

for each isospin multiplet. This multiplicative quantum number is conserved. Thus we have

$$\begin{array}{llll} \pi: & C_n = +1, \quad T = 1 & G = -1 & \\ \rho: & C_n = -1, \quad T = 1 & G = +1 & \rho \to \pi\pi \\ \omega: & C_n = -1, \quad T = 0 & G = -1 & \omega \not\to \pi\pi;\ \omega \to 3\pi \end{array}$$

and

$$\begin{array}{lll} a_1 & T^{C_n}J^P = 1^+1^+ \to \rho\pi; & \not\to \omega\pi \\ b & \phantom{T^{C_n}J^P =}\ 1^-1^+ \to \omega\pi; & \not\to \rho\pi \end{array}$$

and so on.

[The $G$ parity operator is

$$G = Ce^{i\pi T_2}$$

which corresponds to a rotation of $\pi$ about the $y$ axis of (abstract) isospin space. Any state of given $T_3$ is an eigenstate of this operator, for the (integral) isospin multiplets with baryon number, strangeness ... equal to zero, where particle and antiparticle lie in the same multiplet. Thus a $\pi^+$ is rotated into a $\pi^-$ and the $C$ operation then returns a $\pi^+$. A rotation through $\pi$ about any axis in the

$x$–$y$ plane of isospin space will do: the $y$ axis is selected so that all members of a multiplet have the same eigenvalue of $G$, given the conventions already adopted.]

### 11.6 Isospin breaking

The method of calculating relative amplitudes developed in sections 11.3 and 11.4 can be readily extended. It is cumbersome and inelegant, but has advantages. The diagrams shown in figs.11.5 and 11.6 can be regarded as mere book keeping, but they have an obvious physical interpretation in terms of the string picture. The physical origin of the selection rules in the exact cancellation of amplitudes is also obvious. We can see how easy it is to break such a postulated symmetry, and it is easy to trace such effects.

Consider isospin states $\frac{1}{\sqrt{2}}(u\bar{u}+d\bar{d})$, $\frac{1}{\sqrt{2}}(d\bar{d}-u\bar{u})$ such that $\frac{1}{\sqrt{2}}(d\bar{d}-u\bar{u})$ is degenerate with $u\bar{d}$. This requires very special interactions, blind to flavour or with isospin triplet exchange characteristics, AND exact mass degeneracy between $u$ and $d$. Then any infinitesimal mixing interaction takes the degenerate states $u\bar{u}$, $d\bar{d}$ into the isospin eigenstates $\frac{1}{\sqrt{2}}(u\bar{u}+d\bar{d})$, $\frac{1}{\sqrt{2}}(d\bar{d}-u\bar{u})$.

If the $u$ and $d$ masses are not quite equal, the mixed states will be a little bit closer to $u\bar{u}$, $d\bar{d}$. The pair creation amplitudes will not be quite equal either. The $\phi$ is $\approx s\bar{s}$ because $m_s \simeq m_u + 200$ MeV and the mixing is suppressed (Ch.13). Thus $m_u \neq m_d$ breaks isospin, and the electromagnetic interactions also destroy ideal isospin:

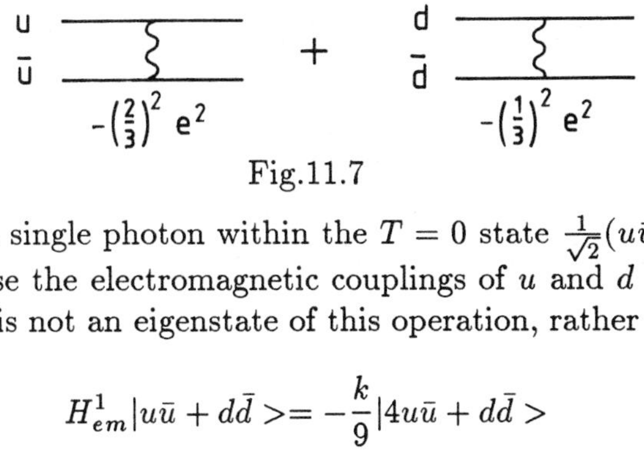


Fig.11.7

The exchange of a single photon within the $T=0$ state $\frac{1}{\sqrt{2}}(u\bar{u}+d\bar{d})$ is shown in fig.11.7. Because the electromagnetic couplings of $u$ and $d$ are different, an isospin eigenstate is not an eigenstate of this operation, rather

$$H^1_{em}|u\bar{u}+d\bar{d}> = -\frac{k}{9}|4u\bar{u}+d\bar{d}> \tag{11.18}$$

Even if $u$ and $d$ had the same mass, the Coulomb interactions would ensure that the mass eigenstates are not pure isospin states.

$$\begin{aligned} H|u\bar{u}+d\bar{d}> &= E_0|u\bar{u}+d\bar{d}> -\frac{k}{9}|4u\bar{u}+d\bar{d}> \\ H|d\bar{d}-u\bar{u}> &= E_1|d\bar{d}-u\bar{u}> -\frac{k}{9}|-4u\bar{u}+d\bar{d}> \end{aligned} \tag{11.19}$$

These equations are diagonalised by mixtures of $u\bar{u}$, $d\bar{d}$ which are not quite the pure isospin states injected on the left hand side.

Isospin in the first quark generation is broken by electromagnetism and by the difference in $u$, $d$ quark masses. In the second and third generations, little trace of an isospin of $(c,s)$, $(t,b)$ remains, although a $(u,d)$ isospin is still present. Thus $D^+(c\bar{d})$, $D^\circ(c\bar{u})$ are a doublet of $(u,d)$ isospin and $D_s^+(c\bar{s})$ is a singlet. The $c - s$ and $t - b$ mass splittings are enormous and have nothing to do with colour or electromagnetism. The old familiar isospin is an approximate result of (low energy) $\alpha_s \gg \alpha$, $m_d - m_u \ll 1$ GeV.

There is also an isospin of the weak interactions, which is very badly broken but in an interesting way. We shall not encounter weak isospin again until Ch.16.

## Problems

11.1 Suppose that an exact alphabetic symmetry exists. Define new alphabetic states

$$\alpha' = \alpha \cos\theta + \beta \sin\theta$$
$$\beta' = -\alpha \sin\theta + \beta \cos\theta$$

Show that the singlet and triplet states $\alpha'\alpha'$, $\frac{1}{\sqrt{2}}(\alpha'\beta' + \beta'\alpha')$ ... defined in this new basis have the same energies as the states $\alpha\alpha$, $\frac{1}{\sqrt{2}}(\alpha\beta + \beta\alpha)$ ...

[For an exact alphabetic symmetry, the Hamiltonian is invariant under arbitrary rotations in alphabetic space.]

11.2* In section 11.3 we asserted that were the representations $\frac{1}{\sqrt{2}}(u\bar{u} + d\bar{d})$ and $\frac{1}{\sqrt{2}}(d\bar{d} - u\bar{u})$ chosen as triplet and singlet respectively, then the vacuum would have to be represented by the composition $d\bar{d} - u\bar{u}$. Convince yourself that this is correct and that the $\rho^\circ - \omega^\circ$ physics really is independent of this convention. [Consider (hypothetical) electromagnetic scattering of $u$ and $d$ quarks and remember that at a scattering vertex an incident quark is annihilated and an outgoing quark created.]

11.3 Use the methods of sections 11.4 and 11.5 to show explicitly that

$$a_1(1^+1^+) \to \rho\pi \qquad \text{but not } \omega\pi$$
$$b_1(1^-1^+) \to \omega\pi \qquad \text{but not } \rho\pi$$

[Notation $T^{C_n} J^P$]

What decay modes do you expect for the states $a_2(1^+2^+)$, $f_2(0^+2^+)$, $h_1(0^-1^+)$ and $f_1(0^+1^+)$?

The $\eta$ and $\eta'$ are both members of the $s$-wave spin singlet $0^-$ nonet, and have isospin zero. Their masses are 549 MeV and 958 MeV respectively. The $\eta$ has a width of 1 KeV and in the language of the Particle Data Group is stable; it does not decay through the strong interactions. The $\eta'$ has a width of 0.25 MeV and does decay strongly. Demonstrate that $\eta$ cannot decay through (isospin conserving) strong interactions and identify the strong decay mode of the $\eta'$.

11.4 The spin triplet $L = 1$ $2^+$ mesons form a nonet with three neutral members, $a_2^\circ$, $f_2^\circ$ and $f_2'$. The $a_2$ is isospin 1, $f_2$ and $f_2'$ are isospin 0. To a good approximation, $f_2$ contains only $u$ and $d$ quarks, while $f_2'$ is $s\bar{s}$. These tensor mesons can be produced in two photon interactions. Calculate the relative rates. [Ignore any mass dependence.]

11.5* The methods of section 11.5 can be extended to particles containing strange quarks. The two $1^+$ nonets (of which $a_1$ and $b_1$ are representative members) contain strange mesons which differ in that one nonet is spin triplet and the other spin singlet. Both strange members can decay into $K^*\pi$ and $\rho K$. Show that the relative sign of the $K^*\pi$ and $\rho K$ amplitudes is opposite for the $C_n = +1$ and $C_n = -1$ members.

The physical strange $1^+$ states are $K_1(1280)$ which decays predominantly into $\rho K$ and $K_1(1400)$ which decays predominantly into $K^*\pi$. Discuss.

# 12. SU(2)→SU(3)$_{colour}$: CHROMODYNAMICS OF HADRONS.

## 12.1 An SU(2) of colour

In this chapter we obtain the properties of gluons coupling to quarks: the properties of SU(3)$_{colour}$ are simply obtained by requiring that quarks come in three colours and imposing a colour independence on the quark-quark interactions. The content and significance is elucidated by considering first the construction and properties of a (non-existent) SU(2) of confined colour.

Suppose we have as many flavours of quark as we like, but that they only come in two colours, say red and green. Require the interaction between quarks to be colour independent, so that

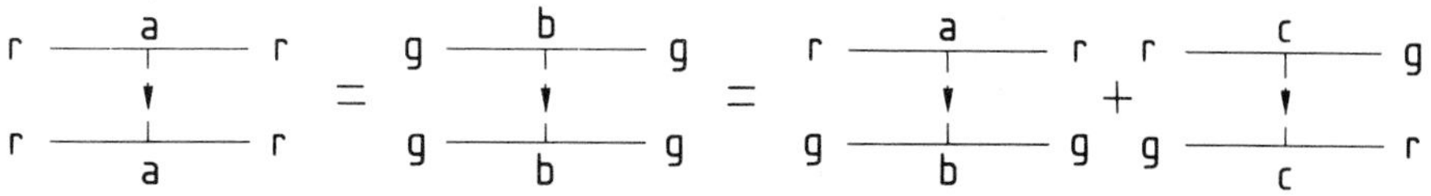


Fig.12.1

As usual, $a^2 = b^2 = ab + c^2$. If there is no field with colour singlet quantum numbers $(r\bar{r} + g\bar{g})$ and the interactions are mediated by a colour triplet, then

$$\begin{aligned} < rr|V|rr > = < gg|V|gg > = \frac{1}{2} < rg + gr|V|rg + gr > = \frac{1}{3}v \\ \frac{1}{2} < rg - gr|V|rg - gr > = -v \end{aligned} \tag{12.1}$$

Note that the singlet state is the unique antisymmetric colour configuration $\left(\frac{1}{\sqrt{2}}\right)\left|\begin{matrix} rg \\ rg \end{matrix}\right|$ and the interaction is attractive in this state if it is repulsive for $(rr)$. Transcription of the alphabet from the previous chapter also yields

$$\begin{aligned} < r\bar{g}|V|r\bar{g} > = \frac{1}{2} < g\bar{g} - r\bar{r}|V|g\bar{g} - r\bar{r} > = < g\bar{r}|V|g\bar{r} > = +\frac{1}{3}\bar{v} \\ \frac{1}{2} < r\bar{r} + g\bar{g}|V|r\bar{r} + g\bar{g} > = -\bar{v} \end{aligned} \tag{12.2}$$

if the interaction is attractive between colour and anticolour. Thus with a repulsive $(rr)$ and attractive $(r\bar{r})$ interaction, characteristic of a vector theory such as electromagnetism, the colour singlet states $\left|\begin{matrix} rg \\ rg \end{matrix}\right|$ and $r\bar{r} + g\bar{g}$ both experience attraction, while the colour triplets are repelled.

$$\begin{aligned} H|singlet > &= (2m + T - v)|singlet > \\ H|triplet > &= (2m + T + \frac{1}{3}v)|triplet > \end{aligned} \tag{12.3}$$

If $m \to \infty$ and $v \to \infty$ in such a way that $2m - v$ remains finite, then only the colour singlets would appear in nature. If the colour fields form tubes of conserved chromoelectric flux this is automatically realised:

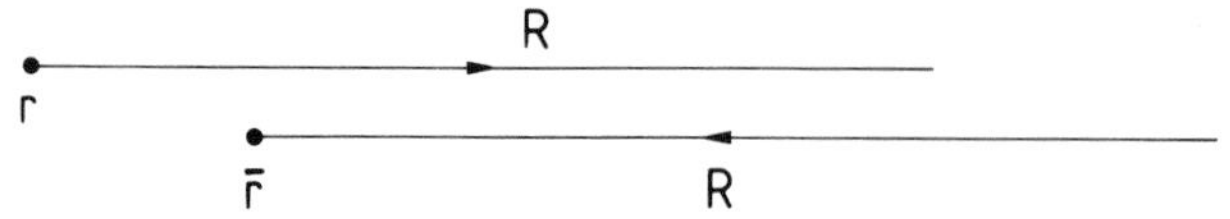


Fig.12.2

Isolated $r, \bar{r}$ are attached to infinitely long colour strings and have infinite mass. If the $R$ field with source $r$ terminates on $\bar{r}$, an infinite length of string has been destroyed, leaving a finite residuum, representing $2m - \bar{v}$.

In such an SU(2) of colour, $g = \bar{r}$ and baryons are $qq$ states (of integral spin). Three quark states are driven to infinite mass.

In a three quark system with only two colours, out of three pairs at least one is symmetric (repulsive interaction) and no more than one pair can be antisymmetric in colour. The lowest energy such state has composition

$$\underset{1}{r}(\underset{23}{rg}-\underset{23}{gr})\frac{1}{\sqrt{2}} \tag{12.4}$$

This is antisymmetric in colour under $2 \leftrightarrow 3$, half symmetric under $1 \leftrightarrow 2 \ldots$ The energy is obtained by sandwiching the colour triplet exchange operator between this state and its conjugate:

$$\begin{aligned} <|V_{12}+V_{23}+V_{13}|> = & \left\langle \frac{\overset{23}{rg}-\overset{23}{gr}}{\sqrt{2}} |V| \frac{\overset{23}{rg}-\overset{23}{gr}}{\sqrt{2}} \right\rangle \\ & + \left\langle \frac{\overset{12}{rr}}{\sqrt{2}} |V| \frac{\overset{12}{rr}}{\sqrt{2}} \right\rangle + \left\langle \frac{\overset{12}{rg}}{\sqrt{2}} |V| \frac{\overset{12}{rg}}{\sqrt{2}} \right\rangle \\ & + \left\langle \frac{\overset{13}{rr}}{\sqrt{2}} |V| \frac{\overset{13}{rr}}{\sqrt{2}} \right\rangle + \left\langle \frac{\overset{13}{rg}}{\sqrt{2}} |V| \frac{\overset{13}{rg}}{\sqrt{2}} \right\rangle \end{aligned} \tag{12.5}$$

(other terms are zero because the spectator colours must match).
Thus

$$<|\Sigma V_{ij}|> = -v + \frac{1}{3}v + <rg|V|rg>$$

Since

$$rg = \frac{1}{2}[(rg+gr)+(rg-gr)],$$

$$<rg|V|rg> = \frac{1}{2}(-1+\frac{1}{3})v = -\frac{1}{3}v \tag{12.6}$$

and the lowest energy $qqq$ configuration acquires $-v$. This is attractive, but not attractive enough, because $-v$ is just enough to bind two quarks. If $2m - v$ is finite as $m$, $v \to \infty$, then $3m - v \to \infty$. For such an SU(2) of confined colour, there are no finite mass $qqq$ states.

But note that the bound $qq$, $q\bar{q}$ states are singlets. With three colours $(r, b, g)$ there is a unique (colour) antisymmetric state of three quarks:

$$\begin{vmatrix} r & b & g \\ r & b & g \\ r & b & g \end{vmatrix}$$

and there is an obvious colour singlet in the $q\bar{q}$ sector: $r\bar{r} + b\bar{b} + g\bar{g}$. If we introduce three colours and again impose colour independence then these three-colour singlets can have finite mass, with all coloured states driven to infinite mass.

The three quark singlet state is totally antisymmetric in colour, so if the quarks are standard fermions the spin-space-flavour structure of the three quark baryons must be totally symmetric. It is.

### 12.2 SU(3) colour coupling coefficients

It all works out quite easily. For colour independence we now require equality of the six amplitudes represented by fig.12.3, and relations such as (12.7).

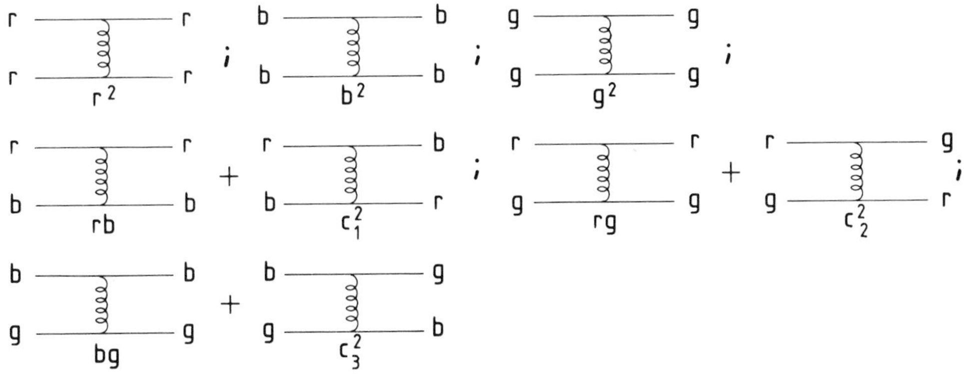


Fig.12.3

$$r^2 = b^2 = g^2 = rb + c_1^2 = rg + c_2^2 = bg + c_3^2 \qquad (12.7)$$

The uninteresting case $c_1 = c_2 = c_3 = 0$ corresponds to colour singlet exchange $r\bar{r}+b\bar{b}+g\bar{g}$. There are eight remaining colour-anticolour combinations, which make up an octet of SU(3)$_{colour}$. There are six like $r\bar{b}$ ...leaving two which must be composed from $r\bar{r}$, $b\bar{b}$, $g\bar{g}$. In fact we cannot solve the coupling equations satisfactorily for only one such state: we have to interpret (12.7) with the replacements

$$r^2 \to r_1^2 + r_2^2, \quad rb \to r_1 b_1 + r_2 b_2 \ \text{ etc.}$$

The two central states must be mutually orthogonal and also orthogonal to the colour singlet $r\bar{r} + b\bar{b} + g\bar{g}$. A popular choice is

$$\frac{1}{\sqrt{2}}(\bar{b}b - \bar{r}r), \quad \frac{1}{\sqrt{6}}(r\bar{r} + b\bar{b} - 2g\bar{g}). \tag{12.8}$$

The first is a member of an $(rb)$ SU(2) triplet, the second an $(rb)$ singlet. This choice is not unique; any pair of orthogonal combinations of these two is equally good for exact SU(3) of colour.

The vertex factors may now be identified from the colour composition of these objects and hence the coupling equations are solved by (12.9)

$$\begin{array}{lll} r_1 = -\frac{1}{\sqrt{2}} & b_1 = \frac{1}{\sqrt{2}} & g_1 = 0 \\ r_2 = \frac{1}{\sqrt{6}} & b_2 = \frac{1}{\sqrt{6}} & g_2 = -\frac{2}{\sqrt{6}} \end{array} \tag{12.9}$$

Then

$$\begin{aligned} r^2 &\to r_1^2 + r_2^2 = b_1^2 + b_2^2 = g_1^2 + g_2^2 = \frac{2}{3} \\ rg &\to r_1 g_1 + r_2 g_2 = -\frac{1}{3} = r_1 b_1 + r_2 b_2 = b_1 g_1 + b_2 g_2 \end{aligned} \tag{12.10}$$

and the coupling equations require $c_1^2 = c_2^2 = c_3^2 = 1$.

The six states with colour composition $rr$, $rg + gr$ ... all pick up $+\frac{2}{3}v$ from the colour octet exchange interaction, while the three antisymmetric states $rg - gr$, ... pick up $-\frac{4}{3}v$. Thus colour octet exchange splits 9 initially degenerate $qq$ states into a 6 and a $\bar{3}$ of colour. Note that a colour combination such as $rg$ is an equal mixture of symmetric and antisymmetric colour states and so picks up $-\frac{1}{3}v$.

The interaction energies of three quark systems can now be worked out. The colour singlet state

$$\begin{vmatrix} r & b & g \\ r & b & g \\ r & b & g \end{vmatrix}$$

is totally antisymmetric in colour. The three pairs of quarks each contribute $-\frac{4}{3}v$ and the mass of a three quark system can be represented by $3m + T - 4v$ where the interaction is attractive if the $rr$ interaction is repulsive. The mean binding interaction is $-\frac{4}{3}v$.

If $3m - 4v$ is finite as $m, v \to \infty$ then $2m - \frac{4}{3}v$ is pushed to $\infty$ (mean binding $-\frac{2}{3}v$) and of course $2m + \frac{2}{3}v \to \infty$. The three quark colour singlet is more deeply bound than any $qq$ state and as $m$, $v \to \infty$ the masses of $q$, $qq$ are driven to infinity. More generally, the masses of all coloured states become infinite. It is already obvious that since the $qqq$ colour singlet is antisymmetric in all pairs, it is the lowest lying $qqq$ system so that all other $qqq$ states $\to \infty$. Increasing the number of quarks does not allow a finite mass until 6 quarks are reached, where the lowest energy configuration has the same mean binding as the $qqq$ colour singlet and so, apart from fine tuning, can fall apart into two baryons. Thus three colours generate three quark, colour singlet, baryons and there are molecular clusters of baryons—nuclei. Nuclear physics is quark chemistry.

Now investigate the meson sector constructed from $q\bar{q}$. The colour singlet is $r\bar{r}+b\bar{b}+g\bar{g}$, and letting colour octet exchange act on one component, say $r\bar{r}$, we obtain

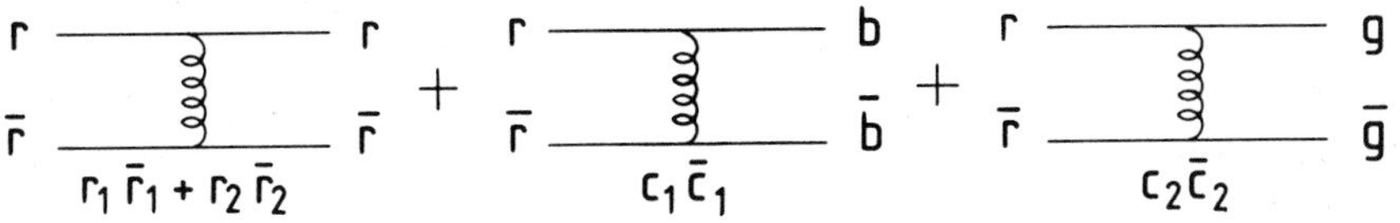


Fig.12.4

Choosing the $r\bar{r}$ force to be attractive ($rr$ was chosen repulsive)

$$r\bar{r} \rightarrow -\frac{2}{3}r\bar{r} - b\bar{b} - g\bar{g} \tag{12.11a}$$

and similarly

$$\begin{aligned} b\bar{b} &\rightarrow -\frac{2}{3}b\bar{b} - r\bar{r} - g\bar{g} \\ g\bar{g} &\rightarrow -\frac{2}{3}g\bar{g} - r\bar{r} - b\bar{g} \end{aligned} \tag{12.11b}$$

so that $r\bar{r}+b\bar{b}+g\bar{g}$ is an eigenstate with interaction energy

$$(-\frac{2}{3}-2)v = -\frac{8}{3}v$$

Then $2m - \frac{8}{3}v$ stays finite as $m, v \rightarrow \infty$ because the mean interaction is again $-\frac{4}{3}v$, as in the $qqq$ colour singlet. Colour octet exchange with long range chromoelectric conserved flux tubes generates finite mass $qqq$ and $q\bar{q}$ states which are colour singlets.

It is easy to extend these arguments to show that $qq\bar{q}$ states are driven to infinite mass ... the next finite mass is the lowest energy configuration of $(q\bar{q})(q\bar{q})$ of course, then $(q\bar{q})(qqq)$. We can have such molecular states in principle and they may exist in $K\bar{K}$, $K^+N$ channels.

Notice that there are three pairs of quarks in a colour singlet baryon, binding $-4v$, and one pair in a $q\bar{q}$ meson, binding $-\frac{8}{3}v$. The colour octet interaction in the colour singlet states is twice as great between $q\bar{q}$ as between $qq$, a result we have encountered before less formally ... it takes $b+g$ to sink a Red field. This result is important.

These colour independent interactions between quark colours, mediated by exchange of an octet of coloured gluons, give for a vector theory (a coloured electromagnetism) the colour singlet configurations of $q\bar{q}$ and $qqq$ as (i) the lowest lying states, and (ii) capable of retaining a finite mass as $m, v \rightarrow \infty$. The string picture embodies these results beautifully and in addition explains the absence of colour polarisation forces—Van der Waal's forces—between nucleons. The colour field cannot be represented by additive potentials.

The structure of QCD is that of a coloured electromagnetism, with conserved colour charges. However, the gluons are coloured and so are self interacting — the theory is non-linear, unlike electromagnetism. It is this self interaction which is supposed to squeeze the colour fields into flux tubes at distances $\gtrsim$ 1fm from the source. There are many theoretical indications that this happens in QCD, and experimentally it must happen, but this mechanism for colour and quark confinement has not been proved from QCD.

### 12.3 Chromomagnetism

Below a distance $\lesssim$ 1fm the interaction is expected to lose its long range flux tube character and to behave like a coloured electromagnetism. The quarks exhibit chromomagnetic moments as well as chromoelectric charges. There is direct evidence for this: the vector mesons $\rho$, $(\omega)$, $K^*$, which are $s$-wave spin triplets, lie considerably higher in mass than the pseudoscalar $\pi$, $K$ which are $s$-wave spin singlets. The spin $\frac{3}{2}$ $s$-wave baryons lie higher than the spin $\frac{1}{2}$ $s$-wave baryons. These splittings are due to colour hyperfine interactions and there is another very pretty example. The isospin 1 hyperon $\Sigma^\circ$ lies higher than the isospin 0 hyperon $\Lambda^\circ$ by 77 MeV, yet they have the same flavour composition, $uds$.

We may lean heavily on an analogous hyperfine interaction in atomic physics, manifest in the 21cm line radiated by atomic hydrogen. This is generated by a spin-flip transition between the (ground) $s$-wave triplet and singlet states. The simplest way to calculate the splitting is to treat the electron cloud as a magnetised sphere: the magnetic induction **B** at the centre of such a sphere (where the proton sits) is proportional to the magnetisation

$$\mathbf{B} = \frac{8\pi}{3}\mathbf{M} \tag{12.12}$$

where **M** is the magnetic moment per unit volume and so is given by

$$\mathbf{M} = \boldsymbol{\mu}_e|\psi(0)|^2. \tag{12.13}$$

The hyperfine interaction in atomic hydrogen is therefore proportional to

$$\boldsymbol{\mu}_p.\mathbf{B} \sim \mu_p\mu_e\mathbf{S}_p.\mathbf{S}_e \qquad \mu_e = \frac{e\hbar}{2mc}, \mu_p = 2.79\frac{e\hbar}{2Mc} \tag{12.14}$$

In the chromomagnetic problem, we are treating very light quarks in a non-relativistic way, mapping a relativistic two body bound state problem onto the Schrödinger equation. Consequently the quark masses that will appear in mass terms, kinetic energy terms and in the chromomagnetic moments will not be the very light current masses ($\sim$ MeV for $u, d$) but effective masses, the constituent masses. There are two major contributions to the generation of constituent masses. First, if a very light fermion is confined within a spherical cavity, the kinetic energy rather than the mass appears in the magnetic moment of the confined fermion. Secondly, transverse string motion effectively contributes to the mass (remember the rotating string) and takes the form of a Lorentz scalar interaction in the Dirac equation. The expectation value of such a term will automatically appear on the bottom line of the expression for magnetic moment, because it is not multiplied by a $\gamma_\mu$ in the Dirac equation. We cannot at present calculate the constituent masses—they must be put in by hand.

If we have a $q$ and $\bar{q}$ on top of each other, the magnetic moment is parallel to the spin for $q$, antiparallel for $\bar{q}$. In a spin triplet, the magnetic moments are antiparallel, but the lowest energy corresponds to moments parallel (parallel currents attract). The same is true for $q\bar{q}$ and $qq$ in colour singlet states: in BOTH these states the colour charges give rise to (chromoelectric) attraction. Then the simplest model we can make for hadron $s$-state masses is

$$M_B = \sum_1^3 m_i + E_{0B} + \sum_{pairs} f_B \left\langle \frac{\mathbf{S}_i.\mathbf{S}_i}{m_i m_j} \right\rangle \tag{12.15a}$$

$$M_M = m_1 + m_2 + E_{0M} + f_M \left\langle \frac{\mathbf{S}_1.\mathbf{S}_2}{m_1 m_2} \right\rangle \tag{12.15b}$$

where the masses are constituent masses. [This is a poor man's version of a famous paper by De Rujula, Georgi and Glashow, *Phys. Rev.* **D12** 147 (1975).]

If we were interested in the mass splitting within an isospin multiplet, we would have to add Coulomb energy terms of the form $q_i q_j \left\langle \frac{1}{r_{ij}} \right\rangle$, and set $m_u \neq m_d$. For our present purposes we set $m_u = m_d = m$ and ignore the Coulomb energy.

We already know the eigenvalues of $\mathbf{S}_1.\mathbf{S}_2$ —

$$\begin{aligned} \mathbf{S}_1.\mathbf{S}_2 &= -\frac{3}{4} \quad \text{for a singlet state} \\ &\phantom{=}\ +\frac{1}{4} \quad \text{for a triplet state} \end{aligned} \tag{12.16}$$

Then among the mesons we have:

$$\begin{aligned} \pi^+(u\bar{d}) \qquad & 140MeV = 2m + E_{0M} - \frac{3}{4}\frac{f_M}{m^2} \\ \rho^+(u\bar{d}) \qquad & 770MeV = 2m + E_{0M} + \frac{1}{4}\frac{f_M}{m^2} \end{aligned} \tag{12.17}$$

from which we obtain at once

$$\frac{f_M}{m^2} = 630 \text{ MeV}$$

$$\begin{aligned} K^+(u\bar{s}) \qquad & 495 \text{ MeV} = m + m_s + E_{0M} - \frac{3}{4}\frac{f_M}{mm_s} \\ K^{*+}(u\bar{s}) \qquad & 892 \text{ MeV} = m + m_s + E_{0M} + \frac{1}{4}\frac{f_M}{mm_s} \end{aligned} \tag{12.18}$$

$$\frac{f_M}{mm_s} = 400 \text{ MeV}$$

and therefore

$$\frac{m_s}{m} = 1.6$$

We can now extract the mass difference between strange and light quarks:

$$\begin{aligned} m_K - m_\pi &= 355 \text{ MeV} = m_s - m - \tfrac{3}{4}(-230) \quad & m_s - m = 180 \text{ MeV} \\ m_K^* - m_\rho &= 122 \text{ MeV} = m_s - m + \tfrac{1}{4}(-230) \quad & m_s - m = 180 \text{ MeV} \end{aligned} \tag{12.19}$$

These relations are not very constrained, but it does work. Let us throw in the $\phi$:

$$M\{\phi(s\bar{s})\} = 1020 \text{ MeV} = 2m_s + E_{0M} + \frac{1}{4}\frac{f_M}{m_s^2} \tag{12.20}$$

and so

$$\begin{aligned} m_\phi - m_\rho = \quad & 250 \text{ MeV} = 2(m_s - m) + \frac{1}{4}\left(\frac{400}{1.6} - 630\right) \\ & = 360 - 95 = 265 \text{ MeV} \end{aligned} \tag{12.21}$$

which is not bad at all. For what it is worth, we may solve the relations

$$\frac{m_s}{m} = 1.6, \quad m_s - m = 180 \text{ MeV}$$

to obtain

$$\begin{aligned} m &= 300 \text{ MeV} \qquad & (\text{constituent } u, d \text{ mass}) \\ m_s &= 480 \text{ MeV} \qquad & (\text{constituent } s \text{ mass}) \end{aligned}$$

All this is not particularly impressive, because there are few constraints. Applying the same techniques to baryons we should find $m_s - m \simeq 180$ MeV, $m_s/m \simeq 1.6$ AND $f_B/m^2 \approx 300$ MeV (because the $qq$ colour coupling is one half the $q\bar{q}$ colour coupling). It works well, as we shall find in the next chaper.

A model of mesons as $q\bar{q}$ pairs held together by the colour exchange forces of QCD, which form confined chromoelectric flux tubes at long range and give rise to chromomagnetic interactions at short range, accounts well for the $s$-wave mesons. We have so far restricted ourselves to the rim states of the $0^-$ and $1^-$ octets: the story is more complicated for the isoscalar mesons and we deal first with the baryons.

## Problems

12.1 Consider the two-colour QCD of section 12.1 in which colour exchange interactions are mediated by a triplet of coloured gluons,

$$\begin{array}{ccc} r\bar{g} & \frac{1}{\sqrt{2}}(g\bar{g} - r\bar{r}) & -\bar{r}g \\ G^+ & G^\circ & G^- \end{array}$$

The relative couplings of quark to gluon required for conservation of colour can be extracted by considering all diagrams of the form of the example given below

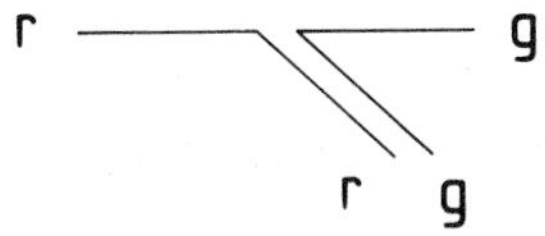


where for conservation of colour both $r\bar{r}$ and $g\bar{g}$ pairs are produced with equal amplitude. Show the standard SU(2) vertex factors

$$\begin{array}{ll} r \leftrightarrow rG^\circ & -\sqrt{\frac{1}{3}} \\ g \leftrightarrow gG^\circ & +\sqrt{\frac{1}{3}} \\ r \leftrightarrow gG^+ & +\sqrt{\frac{2}{3}} \\ g \leftrightarrow rG^- & -\sqrt{\frac{2}{3}} \end{array}$$

are obtained.

Calculate the relative three gluon vertex factors for all six possible three gluon vertices, by creating $r\bar{r} + g\bar{g}$ between colour and anticolour.

12.2 The calculations of problem 1 provide no relation between the quark-gluon and three gluon couplings. Because the gluons carry colour, virtual gluon emission distributes the colour charge on a quark through diagrams such as

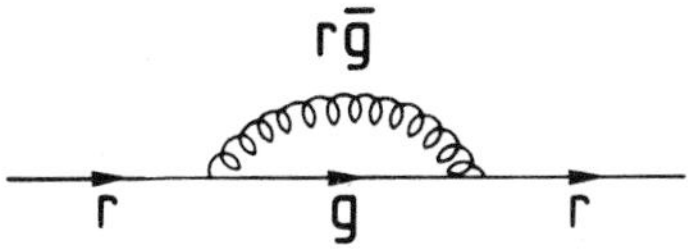


If colour is conserved, the colour charge perceived by a (hypothetical) very soft gluon may not be changed by such virtual loops. [Pion emission and reabsorbtion by a proton does not change the charge perceived by a very soft photon, although at some level this process will contribute to the proton form factor.] Gluons may couple to the intermediate quark or to the virtual gluon in diagrams such as that above. Show that colour conservation in an SU(2) of colour requires the relation between coupling constants

$$G^2 = \frac{8}{3}g^2$$

where $g^2$ is the sum of the squares of the amplitudes for a quark to emit a gluon, and $G^2$ is the sum of the squares of the amplitudes for a gluon to emit a gluon.

[This distribution of colour through the space surrounding a quark has the result that the effective charge perceived by a hard gluon is reduced as the scale probed becomes smaller.]

Another way of obtaining the above result is to consider, say, $G^+ = r\bar{g}$ which couples to colour $g$, and identify the amplitude

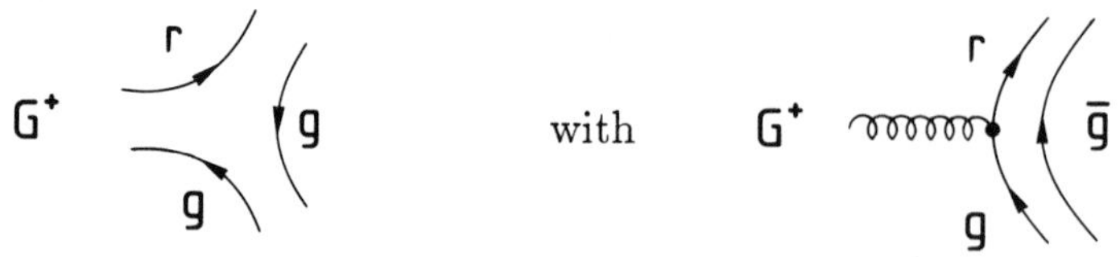


12.3 The two neutral members of the gluon octet are usually represented as

$$\begin{aligned} &\frac{1}{\sqrt{2}}(b\bar{b} - r\bar{r}) \\ &\frac{1}{\sqrt{6}}(r\bar{r} + b\bar{b} - 2g\bar{g}) \end{aligned} \tag{12.8}$$

This is a distant memory of the days when SU(3) was applied in the context of the three flavours $u$, $d$, $s$ and contained a good $(u, d)$ SU(2). If $\mathrm{SU}(3)_{colour}$ is an exact symmetry, any pair of orthonormal combinations of the pair (12.8) is an equivalent representation of the two neutral gluons. Demonstrate that the coupling equations (12.7) are solved by such an arbitrary choice. [There is no such arbitrariness in the choice of the colour singlet configuration.]

12.4 The mass differences within the $T = 1$ and $T = \frac{1}{2}$ $0^-$ multiplets are

$$\begin{aligned} m_{\pi^\pm} - m_{\pi^\circ} &= 4.60 \text{ MeV} \\ m_{K^\circ} - m_{K^\pm} &= 4.0 \text{ MeV} \end{aligned}$$

Extend (12.15a) to allow for different $u$ and $d$ quark constituent masses and include quark-antiquark coulomb interactions. Deduce what you can about the value of the quantity

$$m_d - m_u$$

and the size of the $0^-$ mesons.

12.5 Calculate the hyperfine splitting in the ground state of atomic hydrogen [eqs. (12.12–14)] and show that the corresponding (radio) frequency is 1420 MHz, a wavelength of 21 cm.

Estimate the lifetime for spontaneous decay of the upper state.

[It is interesting that were particle magnetic moments due to pairs of monopoles, rather than current loops, the wavelength would be 42 cm.]

Given the observed hyperfine splitting in the $s$-wave mesons, estimate the value of the dimensionless quantity $\alpha_s$ (which corresponds to $e^2/\hbar c$ in electromagnetism) for the colour fields.

# 13. CHROMODYNAMICS OF HADRONS—GROUND STATE BARYONS AND ISOSCALAR MESONS.

## 13.1 Baryon hyperfine splitting

The $s$-wave baryons composed of the light quarks $u$, $d$, $s$ comprise the spin $\frac{3}{2}^+$ $(SU(3)_{flavour})$ decuplet and the $\frac{1}{2}^+$ octet:

| | | | | | | | | |
|---|---|---|---|---|---|---|---|---|
| 938 MeV | | $p_{\bullet uud}$ $n_{\bullet udd}$ | | $\Delta^{++}_{\bullet uuu}$ | $\Delta^{+}_{\bullet uud}$ | $\Delta^{0}_{\bullet udd}$ | $\Delta^{-}_{\bullet ddd}$ | 1232 MeV |
| 1190 | $\Sigma^{+}_{\bullet uus}$ | $\Sigma^{0}_{\bullet uds}$ | $\Sigma^{-}_{\bullet dds}$ | | $\Sigma^{+}_{\bullet uus}$ | $\Sigma^{0}_{\bullet uds}$ | $\Sigma^{-}_{\bullet dds}$ | 1385 |
| 1115 | | $\Lambda^{0}_{\bullet}$ | | | $\Xi^{0}_{\bullet uss}$ | $\Xi^{-}_{\bullet dss}$ | | 1530 |
| 1320 | | $\Xi^{0}_{\bullet uss}$ $\Xi^{-}_{\bullet dss}$ | | | | $\Omega^{-}_{\bullet sss}$ | | 1672 |

The decuplet is approximately equally spaced with increasing strangeness and the splitting is easily understood in terms of an $s$ quark $\approx 150$ MeV heavier than $u$ and $d$. The octet is certainly not equally spaced and $\Sigma^\circ$ and $\Lambda^\circ$ are split. All these features are quantitatively accounted for in terms of chromomagnetic interactions. Our simple model for baryon mass is

$$M = \sum m_i + E_{0B} + E_{HS} \tag{13.1}$$

where

$$E_{HS} = \sum_{i \neq j} f_B \frac{< \mathbf{S}_i.\mathbf{S}_j >}{m_i m_j}$$

and we expect to find

$$f_B \approx 300 \text{ MeV}.$$

In the decuplet, all quark pairs are symmetric in spin and so we know that $< \mathbf{S}_i.\mathbf{S}_j >= +\frac{1}{4}$ for all three pairs. The octet is more complicated. Remember that because the colour singlet is totally antisymmetric and the space part is $s$-wave for all pairs, the spin functions must be symmetric under interchange of any identical pair. We may represent the spins in the $\Delta^+$ by

$$\Delta^+ \quad J_z = +\frac{3}{2} \qquad \underset{uud}{\uparrow\uparrow\uparrow} \tag{13.2}$$

Lowering each spin in turn, we obtain

$$\Delta^+ \quad J_z = +\frac{1}{2} \qquad \underset{uud}{\uparrow\uparrow\downarrow} + \underset{uud}{\uparrow\downarrow\uparrow} + \underset{uud}{\downarrow\uparrow\uparrow} \quad \left(\times\frac{1}{\sqrt{3}}\right) \tag{13.3}$$

We can at once write down a state made from $uud$ with $J_z = +\frac{1}{2}$ which is symmetric under interchange of the two $u$ quarks and is orthogonal to $\Delta^+(J_z = +\frac{1}{2})$:

$$p \quad J_z = +\frac{1}{2} \qquad 2\underset{uud}{\uparrow\uparrow\downarrow} - (\underset{uu}{\uparrow\downarrow} + \underset{uu}{\downarrow\uparrow})\underset{d}{\uparrow} \quad \left(\times\frac{1}{\sqrt{6}}\right) \tag{13.4}$$

All the rim states of the octet have two quarks identical, so this structure applies to them all. In order to obtain the hyperfine splitting we must evaluate terms such as

$$< p|\mathbf{S}_i.\mathbf{S}_j|p > \tag{13.5}$$

for all three pairs of quarks. This is not difficult.

The first pair in (13.4) is symmetric in spin so $<\mathbf{S}_1.\mathbf{S}_2>=+\frac{1}{4}$. For the pair 2 and 3

$$2\uparrow\uparrow\downarrow-(\uparrow\downarrow+\downarrow\uparrow)\uparrow \;=\; \uparrow[2\uparrow\downarrow-\downarrow\uparrow]-\downarrow[\uparrow\uparrow] \tag{13.6}$$

so

$$\begin{aligned}<\mathbf{S}_2.\mathbf{S}_3>=\frac{1}{6}<2\uparrow\downarrow-\downarrow\uparrow|\mathbf{S}_2.\mathbf{S}_3|2\uparrow\downarrow-\downarrow\uparrow>+\frac{1}{6}<\uparrow\uparrow|\mathbf{S}_2.\mathbf{S}_3|\uparrow\uparrow>\\ <=+\frac{1}{4}>\end{aligned} \tag{13.7}$$

$$\begin{aligned}
\mathbf{S}_2.\mathbf{S}_3|\uparrow\downarrow+\downarrow\uparrow> &= \frac{1}{4}|\uparrow\downarrow+\downarrow\uparrow>\\
\mathbf{S}_2.\mathbf{S}_3|\uparrow\downarrow-\downarrow\uparrow> &= -\frac{3}{4}|\uparrow\downarrow-\downarrow\uparrow>\\
\therefore\quad \mathbf{S}_2.\mathbf{S}_3|\uparrow\downarrow> &= |-\frac{1}{4}\uparrow\downarrow+\frac{1}{2}\downarrow\uparrow>\\
\mathbf{S}_2.\mathbf{S}_3|\downarrow\uparrow> &= |\frac{1}{2}\uparrow\downarrow-\frac{1}{4}\downarrow\uparrow>\\
\mathbf{S}_2.\mathbf{S}_3|2\uparrow\downarrow-\downarrow\uparrow> &= |-\uparrow\downarrow+\frac{5}{4}\downarrow\uparrow>
\end{aligned}$$

$$<\mathbf{S}_2.\mathbf{S}_3>_8=\frac{1}{6}[-2-\frac{5}{4}]\;+\;\frac{1}{6}[+\frac{1}{4}]\;=\;-\frac{1}{2} \tag{13.8}$$

$<\mathbf{S}_1.\mathbf{S}_3>_8$ has the same value. The hyperfine interaction in the octet rim states therefore contributes

$$+\frac{1}{4}\frac{f_B}{m^2}-\frac{f_B}{mm'} \tag{13.9}$$

where $m$ is the mass of the identical quarks and $m'$ is the mass of the odd quark. We can now write down expressions for the masses of the RIM octet states and the decuplet states:

| | Octet | | Decuplet |
|---|---|---|---|
| $N$ | $3m+E_{0B}-\frac{3}{4}\frac{f_B}{m^2}$ | $\Delta$ | $3m+E_{0B}+\frac{3}{4}\frac{f_B}{m^2}$ |
| $\Sigma$ | $2m+m_s+E_{0B}+\frac{1}{4}\frac{f_B}{m^2}-\frac{f_B}{mm_s}$ | $\Sigma$ | $2m+m_s+E_{0B}+\frac{1}{4}\frac{f_B}{m^2}+\frac{1}{2}\frac{f_B}{mm_s}$ |
| $\Xi$ | $m+2m_s+E_{0B}+\frac{1}{4}\frac{f_B}{m_s^2}-\frac{f_B}{mm_s}$ | $\Xi$ | $m+2m_s+E_{0B}+\frac{1}{2}\frac{f_B}{mm_s}+\frac{1}{4}\frac{f_B}{m_s^2}$ |
| | | $\Omega$ | $3m_s+E_{0B}+\frac{3}{4}\frac{f_B}{m_s^2}$ |

and obtain

$$\begin{aligned}
\Delta-N &= 294\text{ MeV}=\tfrac{3}{2}\tfrac{f_B}{m^2} & \tfrac{f_B}{m^2}&\simeq 200\text{ MeV}\\
\Sigma_{10}-\Sigma_8 &= 195\text{ MeV}=\tfrac{3}{2}\tfrac{f_b}{mm_s} & \tfrac{m_s}{m}&\simeq 1.5\\
\Xi_{10}-\Xi_8 &= 210\text{ MeV}=\tfrac{3}{2}\tfrac{f_B}{mm_s} & \tfrac{m_s}{m}&\simeq 1.5
\end{aligned}$$

These results are in remarkably good agreement with the results of the same model applied to mesons. We obtain almost exactly the same value for $m_s/m$

in both cases. The quantity $f_B$ is $\approx f_M/3$ rather than $f_M/2$ but this is not unexpected. The baryons contain three quarks, the short range $qq$ forces are weaker and we expect baryons to be a bit bigger and the quark wave functions less dense at the origin. Elastic scattering of (charged) pions from electrons has shown that the pion is a bit smaller than the nucleon:

$$< r_\pi^2 >^{\frac{1}{2}} = 0.66 \text{ fm} \qquad < r_p^2 >^{\frac{1}{2}} = 0.8 \text{ fm}$$

The chromomagnetic splitting is diluted by the cube of the radius and a 20% difference in the effective radius entering the expression for the hyperfine splitting is more than enough to account for $\frac{f_B}{m^2} \lesssim 200$ MeV rather than $\sim 300$ MeV.

Comparing masses within the octet

$$\Sigma - N = m_s - m + \frac{f_B}{m^2} - \frac{f_B}{mm_s}$$

$$252 \text{ MeV} = m_s - m + 67 \text{ MeV} \qquad m_s - m \simeq 185 \text{ MeV}$$

$$\Xi - \Sigma = m_s - m + \frac{1}{4}\frac{f_B}{m_s^2} - \frac{1}{4}\frac{f_B}{m^2}$$

$$130 \text{ MeV} = m_s - m + 22 - 50 \text{ MeV} \qquad m_s - m \simeq 160 \text{ MeV}$$

The decuplet masses work out quite well too, and the whole treatment can be improved by including a piece in $E_0$ which is proportional to $\Sigma(1/m_i)$, corresponding to kinetic energy $\sim \Sigma(< p^2 > /2m_i)$.

Finally we turn our attention to the central members of the octet, $\Sigma^\circ$ and $\Lambda^\circ$. The quark composition of these two is the same. The spin structure of $\Sigma^\circ$ is that of $\Sigma^+$, replacing $uu$ by $ud$:

$$\Sigma^\circ = \frac{1}{\sqrt{6}} \left\{ \begin{matrix} 2\uparrow\uparrow\downarrow & -( \uparrow\downarrow & + \downarrow\uparrow ) & \uparrow \\ uds & ud & ud & s \end{matrix} \right\} \tag{13.10}$$

It is degenerate with $\Sigma^+$, $\Sigma^-$ (ignoring the $u-d$ mass difference and the Coulomb energy). The $\Lambda^\circ$ spin structure is orthogonal to both $\Sigma^\circ$ and the decuplet $uds$ and is obviously

$$\Lambda^\circ = \frac{1}{\sqrt{2}} \left\{ \begin{matrix} ( \uparrow\downarrow & - \downarrow\uparrow ) & \uparrow \\ ud & ud & s \end{matrix} \right\} \tag{13.11}$$

In $\Sigma^\circ$ the $u$ and $d$ quarks couple to give spin 1, in $\Lambda^\circ$ the $u$ and $d$ spins couple to spin zero.

The quark spin structure in $\Lambda^\circ$ gives chromomagnetic splittings equivalent to one antisymmetric pair $(ud)$, one symmetric pair and one odd pair.

$$E_{HS}(\Lambda^\circ) = -\frac{3}{4}\frac{f_B}{m^2} + \frac{1}{4}\frac{f_B}{mm_s} - \frac{1}{4}\frac{f_B}{mm_s}$$

Then

$$\begin{aligned} \Sigma^\circ - \Lambda^\circ = 77 \text{ MeV} &= \frac{1}{4}\frac{f_B}{m^2} - \frac{f_B}{mm_s} + \frac{3}{4}\frac{f_B}{m^2} \\ &= 67 \text{ MeV } (m_s/m = 1.5) \\ &\phantom{=} 75 \text{ MeV } (m_s/m = 1.6) \end{aligned}$$

Thus we have a rather good understanding of the ground state baryons as well as the ground state mesons with an (approximately) common set of parameters. Note that as $m_s \to m$ all members of the baryon octet become degenerate, and all members of the baryon decuplet become degenerate. $\mathrm{SU}(3)_{flavour}$ is broken by the mass of the strange quark.

In obtaining these impressive results we assumed that the hadrons are colour singlets and that the wave functions of identical quarks must be totally antisymmetric—the Pauli exclusion principle. The interactions were taken to be those of an SU(3) of colour and the colour field a vector field. Because the three quark colour singlet is antisymmetric in colour, the spin functions of identical quarks in the baryon octet and decuplet are symmetric. There is a further check that can be applied—the magnetic moments of the octet baryons.

### 13.2 Baryon magnetic moments

Assume that the bound quarks behave like Dirac particles with the usual fractional charges and constituent masses. The quark magnetic moments have values

$$
\begin{aligned}
<\uparrow |\hat{\mu}| \uparrow> &= \frac{e_q \hbar}{2 m_q c} \\
<\downarrow |\hat{\mu}| \downarrow> &= -\frac{e_q \hbar}{2 m_q c}
\end{aligned}
\qquad (13.12)
$$

For the rim octet states we have

$$
\begin{aligned}
\mu &= \frac{1}{6} \quad < 2 \underset{aab}{\uparrow\uparrow\downarrow} -( \underset{aa}{\uparrow\downarrow} + \underset{aa}{\downarrow\uparrow} ) \underset{b}{\uparrow} |\hat{\mu}| 2 \underset{aab}{\uparrow\uparrow\downarrow} -( \underset{aa}{\uparrow\downarrow} + \underset{aa}{\downarrow\uparrow} ) \underset{b}{\uparrow} > \\
&= \frac{1}{6}\{4(\mu_a + \mu_a - \mu_b) + 2\mu_b\} \\
&= \frac{4}{3}\mu_a - \frac{1}{3}\mu_b
\end{aligned}
\qquad (13.13)
$$

The magnetic moments can be written down at once. For the proton the two identical quarks are $u$ and we have

$$a = u \qquad b = d$$

$$\mu_p = \left(\frac{4}{3}\frac{2}{3} + \frac{1}{3}\frac{1}{3}\right)\frac{e\hbar}{2mc} = \frac{e\hbar}{2mc} = 2.79\frac{e\hbar}{2M_p c} \qquad (13.14)$$

Measure the magnetic moments in nuclear magnetons. Then

$$
\begin{array}{llllll}
n & a = d & b = u & \mu_n = \left(\frac{4}{3}\left(-\frac{1}{3}\right) - \frac{1}{3}\frac{2}{3}\right) 2.79 = -\frac{2}{3}\mu_p & = -1.86 \; (-1.91) \\
\Sigma^+ & a = u & b = s & \mu_{\Sigma^+} = \left(\frac{4}{3}\frac{2}{3} + \frac{1}{3}\frac{1}{3}\frac{m}{m_s}\right) 2.79 & = +2.69 \; (+2.42 \pm 0.05) \\
\Sigma^- & a = d & b = s & \mu_{\Sigma^-} = \left(\frac{4}{3}\left(-\frac{1}{3}\right) + \frac{1}{3}\frac{1}{3}\frac{m}{m_s}\right) 2.79 & = -1.03 \; (-1.16 \pm 0.03) \\
\Xi^\circ & a = s & b = u & \mu_{\Xi^\circ} = \left(\frac{4}{3}\left(-\frac{1}{3}\right)\frac{m}{m_s} - \frac{1}{3}\frac{2}{3}\right) 2.79 & = -1.45 \; (-1.25 \pm 0.02) \\
\Xi^- & a = s & b = d & \mu_{\Xi^-} = \left(\frac{4}{3}\left(-\frac{1}{3}\right)\frac{m}{m_s} + \frac{1}{3}\frac{1}{3}\right) 2.79 & = -0.52 \; (-0.65 \pm 0.02)
\end{array}
$$

(The experimental values are given in brackets.)

$\Lambda^\circ$ is very easy to calculate. The spin structure is

$$\underset{ud}{(\uparrow\downarrow} - \underset{ud}{\downarrow\uparrow)}\underset{s}{\uparrow}$$

with $u$ and $d$ coupling to spin zero. Then

$$\mu_{\Lambda^\circ} = \mu_s = -\frac{1}{3}\frac{m}{m_s}2.79 = -0.62\ (-0.61)$$

with $m_s/m = 1.5$.

Overall the agreement, which is not perfect, is remarkably good. We may also note that from the normalising condition

$$\frac{e\hbar}{2mc} = \frac{2.79e\hbar}{2M_pc}$$

we obtain

$$m \simeq 330 \text{ MeV}$$

which is in good agreement with our earlier estimate of the $u$, $d$ constituent mass obtained at the end of chapter 12.

The model we have employed for the ground state hadrons is extremely crude and it would be remarkable if the mass splittings and magnetic moments were in any better accord with the data. It is worth recapitulating the achievements of this crude picture. First, one would expect the lowest lying meson and baryon states to be $s$-waves. This does not violate the Pauli principle provided that quarks come in three colours. With colour exchange interactions generating an SU(3) of colour the lowest lying states are colour singlets and with confined colour we have reason to expect that all physical states are colour singlets. The three quark colour singlet is totally antisymmetric and so the baryons must be spin-space (flavour) symmetric, and this is directly verified by the octet baryon magnetic moments. With vector colour fields, $q\bar{q}$ and $qqq$ states are bound, and with the colour couplings appropriate to colour singlet states, the correct pattern of colour hyperfine interactions is generated in both meson and baryon sectors.

The representation of bound states of highly relativistic quarks in terms of a non-relativistic reduction required the introduction of the effective, constituent, masses. While some justification for this phenomenological guess can be obtained from studying states of free (relativistic) quarks confined within a cavity, approximately the same constituent masses are obtained from both meson and baryon masses, and independently from the baryon magnetic moments. Constituent masses work, and this is the best justification. The properties of hadrons require for their understanding most of the postulated features of QCD.

### 13.3 Isosinglet states, annihilation and flavour mixing

One further aspect of the chromodynamics of ground state hadrons is encountered only in the neutral meson states. The $u$, $d$ and $s$ masses are all different and one might expect the neutral meson eigenstates to be $u\bar{u}$, $d\bar{d}$, $s\bar{s}$ ....

The colour potentials do not mix flavours, but there are annihilation diagrams which do:

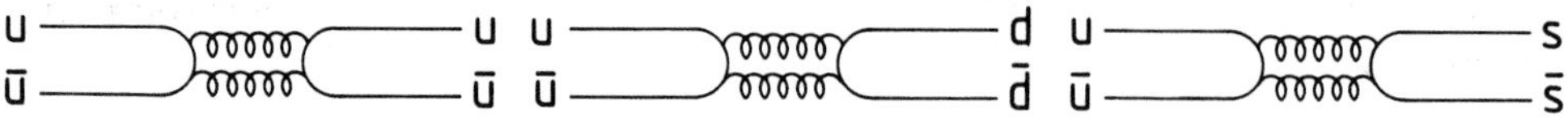


Fig.13.1

Since gluons are coloured, at least two are needed to connect colourless $q\bar{q}$ states. Since the gluon has $C = -1$, $C = +1$ states annihilate through two gluon channels and $C = -1$ states through three gluon channels.

Had we only $u$ and $d$ quarks

$$H|u\bar{u}> = E_u|u\bar{u}> + A|u\bar{u}> + A|d\bar{d}> \tag{13.15a}$$

$$H|d\bar{d}> = E_d|d\bar{d}> + A|d\bar{d}> + A|u\bar{u}> \tag{13.15b}$$

assuming the annihilation amplitudes are flavour independent. If $E_u = E_d$ then the eigenstates are easily found by adding and subtracting eqs.(13.15)

$$\begin{aligned} H|d\bar{d} - u\bar{u}> &= E|d\bar{d} - u\bar{u}> \\ H|u\bar{u} + d\bar{d}> &= (E + 2A)|u\bar{u} + d\bar{d}> \end{aligned} \tag{13.16}$$

The energy of the isospin triplet state is not shifted by the annihilation term—it remains degenerate with $u\bar{d}$, $\bar{u}d$. The isospin singlet state is shifted by $2A$.

In fact $E_d - E_u \sim 2(m_d - m_u) \sim 12$ MeV, but the isospin states will be a very good approximation to the eigenstates provided that $2A \gg E_d - E_u$.

Let $E_d \simeq E_u = E$ and couple in $s\bar{s}$ pairs. There is no reason to suppose that the annihilation amplitudes will be the same for $u\bar{u} \to u\bar{u}$, $u\bar{u} \to s\bar{s}$, $s\bar{s} \to s\bar{s}$, but it is plausible that they factorise†. Then

$$\begin{aligned} H|u\bar{u}> &= E|u\bar{u}> + \alpha^2|u\bar{u}> + \alpha^2|d\bar{d}> + \alpha\beta|s\bar{s}> \\ H|d\bar{d}> &= E|d\bar{d}> + \alpha^2|u\bar{u}> + \alpha^2|d\bar{d}> + \alpha\beta|s\bar{s}> \\ H|s\bar{s}> &= E_s|s\bar{s}> + \beta^2|s\bar{s}> + \alpha\beta|u\bar{u}> + \alpha\beta|d\bar{d}> \end{aligned} \tag{13.17}$$

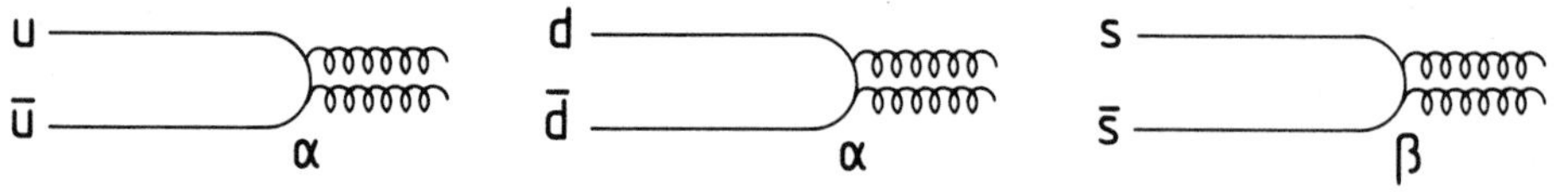


Fig.13.2

Subtract the first two equations of (13.17): $(d\bar{d} - u\bar{u})$ is an eigenstate with energy $E$, unshifted by the annihilation terms.

Add the first two equations and we then have two coupled equations for the two isosinglet states:

$$\begin{aligned} H|u\bar{u} + d\bar{d}> &= (E + 2\alpha^2)|u\bar{u} + d\bar{d}> + 2\alpha\beta|s\bar{s}> \\ H|s\bar{s}> &= (E_s + \beta^2)|s\bar{s}> + \alpha\beta|u\bar{u} + d\bar{d}> \end{aligned} \tag{13.18}$$

---

† This factorisation obtains in an explicit calculation with quarks confined within a spherical cavity; Donaghue and Gomm *Phys. Rev.* **D28** 2800 (1983)

The annihilation terms are all small except in the $J^{PC} = 0^{-+}$ system. In the ($s$-wave) vectors $J^{PC} = 1^{--}$ and three gluons are required. The vector mesons $\rho^\circ$ and $\omega^\circ$ are almost degenerate in mass and $\rho^\circ \simeq \frac{1}{\sqrt{2}}(d\bar{d} - u\bar{u})$, $\omega^\circ \simeq \frac{1}{\sqrt{2}}(u\bar{u} + d\bar{d})$, $\phi \simeq s\bar{s}$. In $s$-wave $0^{-+}$ $q\bar{q}$ systems, which can annihilate through two gluons, we find

$$\begin{array}{ll} \pi^\circ & 136 \text{ MeV} \\ \eta & 549 \text{ MeV} \\ \eta' & 958 \text{ MeV} \end{array}$$

The $\pi^\circ$ is approximately degenerate with $\pi^\pm$, and does not contain strange quarks. Both the isosinglets $\eta$ and $\eta'$ contain strange quarks, and we know that $\eta'$ has the greater proportion; $\eta'$ contains only half as much $u\bar{u} + d\bar{d}$ as $\eta$. The reason is that

$$\sigma(\pi^- p \to \eta' n)/\sigma(\pi^- p \to \eta n) \simeq 0.5$$

For both processes the quark exchange diagram is

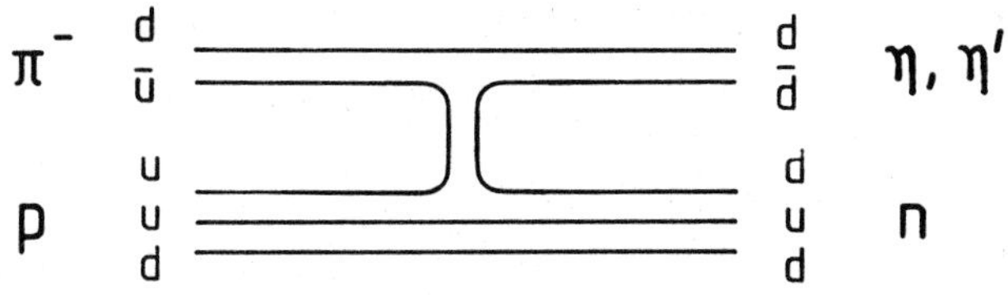


Fig.13.3

Then $|<\eta'|d\bar{d}>|^2 \simeq \frac{1}{2}|<\eta|d\bar{d}>|^2$ and the normalised orthogonal states are

$$\begin{aligned} \eta &= \frac{1}{\sqrt{3}}(u\bar{u} + d\bar{d} - s\bar{s}) \\ \eta' &= \frac{1}{\sqrt{6}}(u\bar{u} + d\bar{d} + 2s\bar{s}) \end{aligned} \qquad (13.19)$$

(where we have chosen signs to ensure that $M_{\eta'} > M_\eta$).

The quark composition, obtained from the relative production rates, is an important constraint, because $\alpha$ and $\beta$ (13.18) determine the masses of $\eta$, $\eta'$ and the quark composition. However, using the above quark composition (13.19) we can solve for $\alpha$ and $\beta$ and a consistency check is built in.

$$\begin{aligned} M_\eta|u\bar{u} + d\bar{d} - s\bar{s}> &= H|u\bar{u} + d\bar{d} - s\bar{s}> \\ &= (E + 2\alpha^2 - \alpha\beta)|u\bar{u} + d\bar{d}> -(E_s + \beta^2 - 2\alpha\beta)|s\bar{s}> \end{aligned} \qquad (13.20)$$

Therefore

$$M_\eta = E + 2\alpha^2 - \alpha\beta = E_s + \beta^2 - 2\alpha\beta$$

$$\text{Similarly} \quad M_{\eta'} = E + 2\alpha^2 + 2\alpha\beta = E_s + \beta^2 + \alpha\beta$$

The right hand pair of equations is consistent; $\alpha$ and $\beta$ can be obtained from the left hand pair alone. The quantity $E$ is the mass of $(d\bar{d} - u\bar{u})$ (13.16, 17) and so is the pion mass. Then

$$M_{\eta'} - M_\eta = 410 \text{ MeV} = 3\alpha\beta$$

$$M_{\eta'} + 2M_\eta - 3m_\pi = 1648 \text{ MeV} = 6\alpha^2$$

$$\alpha^2 = 275 \text{ MeV} \qquad \beta^2 = 68 \text{ MeV}$$

From the right hand pair of equations

$$E_s - E = 2\alpha^2 - \beta^2 + \alpha\beta = 618 \text{ MeV}$$

We already know $E_s - E$ independently, since

$$E_s - E = 2(m_s - m_u) - \frac{3}{4}\left(\frac{f_M}{m_s^2} - \frac{f_M}{m^2}\right)$$

and $m_s - m_u \simeq 180$ MeV, $f_M/m^2 = 630$ MeV, $m_s/m \simeq 1.5$ from the pseudoscalar and vector octet rim states. Using these values

$$E_s - E = 2(180) + \frac{3}{4}630\left(1 - \frac{1}{1.5^2}\right) = 622 \text{ MeV}$$

The whole picture is beautifully consistent, and the factorisation hypothesis is valid.

It is only in $s$-wave spin singlets that the annihilation terms are important. When the orbital angular momentum is $> 0$, $q$ and $\bar{q}$ are kept apart by the centrifugal barrier and annihilation is inhibited. The $s$-wave spin triplets require a minimal configuration of three gluons and this is evidently suppressed. Radial excitations of the $0^-$ states ($s$-waves but not ground state $s$-waves) should exhibit an intermediate degree of mixing, but data are sparse.

For substantial mixing of two states, the mixing term must be comparable with the energy difference between the two states. It is clear that even for ground state $0^-$ mesons, the $c\bar{c}$ and $b\bar{b}$ structures will be very pure (masses $\sim 3$ GeV and 10 GeV respectively) and will scarcely mix with each other or with the light quark states. These mesons, hidden charm and hidden bottom states, provide a beautiful laboratory for testing our ideas about meson structure—the Bohr atoms of QCD. They are discussed in the next chapter.

### 13.4 The spreading of colour charge

The effective colour charge of a quark (or a gluon) is not localised on the particle, in contrast to the electric charge on an electron. A blue quark can interact with an (anti-)blue antiquark by exchange of a $r\bar{b}$ gluon: this is just a representation of the colour exchange forces acting once in perturbation theory. The colour potential surrounding a quark itself carries colour and a probe sensitive to blue will find that charge smeared out round a blue quark through emission of virtual gluons carrying blue. Thus a hard gluon capable of resolving short distances will experience an effective charge; the (conserved) blue diluted by a form factor. Since colour is confined, it is not possible to define colour charges absolutely by probing with an infinitely soft gluon. The colour coupling has to be defined at a particular value of momentum transfer, $q^2$, and the effective colour charges fall slowly as $q^2$ increases.

In a deeply bound state of massive quark-antiquark the pair are deeply immersed in the colour field and again the effective colour charge is reduced. The dimensionless colour coupling $\alpha_s$ in fact falls logarithmically as $q^2$ increases. As $q^2$ decreases the non-perturbative region is entered and there exist at present no reliable calculations in this region. The conjecture is that confining flux tubes of colour develop at long range.

## Problems

13.1 Modify the mass relations for the $s$-wave octet baryons given below (13.9) to allow for the difference in mass between $u$ and $d$ quarks and for the coulomb interactions. Extract the mass difference

$$m_d - m_u$$

and infer what you can about the size of the baryons. Compare your results with those of problem 12.4

[Relevant mass differences are

| | |
|---|---|
| $n - p$ | 1.293 MeV |
| $\Sigma^- - \Sigma^+$ | 7.97 MeV |
| $\Xi^- - \Xi^\circ$ | 6.4 MeV ] |

13.2 Consider the vector mesons $\rho^\circ$ and $\omega^\circ$, which have masses and widths 772, 153; 783, 9.8 MeV respectively. The $\omega^\circ$ decays predominantly into $\pi^+\pi^-\pi^\circ$ but has a branching ratio of 0.017 into $\pi^+\pi^-$ (and 0.09 into $\pi^\circ\gamma$). The decay into $\pi^+\pi^-$ is not consistent with $C = -1$ and isospin zero. Use the quark model to estimate the proportion of $I = 1$ in the $\omega^\circ$. Relevant effects are the $d-u$ mass difference, coulomb interactions, and annihilation through a single virtual photon.

Consider the patterns of $\rho - \omega$ interference in the final state $\pi^+\pi^-$.

[Assume that $\rho^\circ$ and $\omega^\circ$ have negligible strange quark content.]

13.3(i) Calculate the $\Omega^-$ magnetic moment.

(ii) In section 13.2 we calculated the static magnetic moments of the octet baryons, with the exception of the $\Sigma^\circ$ which has a lifetime of $5.8(\pm 1.3) \times 10^{-20}$s. Calculate the transition matrix element $< \Sigma^\circ|\hat{\mu}|\Lambda^\circ >$ which governs the M1 decay $\Sigma^\circ \to \Lambda^\circ\gamma$. Estimate the lifetime of $\Sigma^\circ$. How could the square of the matrix element be measured?

(iii) Show that the electromagnetic transition

$$\Sigma^-(1385) \to \Sigma^-\gamma$$

is forbidden, whereas

$$\Sigma^+(1385) \to \Sigma^+\gamma$$

is not forbidden.

How could this quark model prediction be checked?

13.4 We have cheerfully asserted that a (neutral) meson which is an eigenstate of $C$ with $C = -1$ cannot convert into a single gluon (because of colour) and cannot convert into two gluons because of conservation of $C$. It is fairly obvious that a $C = -1$ state cannot convert into two $C = -1$ states, but of the eight gluons only two have $C = -1$; the $C$ operation transforms $r\bar{b}$ into $\bar{r}b$. The methods of section 11.2 can be extended to SU(3) of colour: between quark and antiquark create

$$r\bar{r} + \bar{r}r + b\bar{b} + \bar{b}b + g\bar{g} + \bar{g}g.$$

A colour singlet $C = -1$ meson is to be represented by

$$r\bar{r} - \bar{r}r + b\bar{b} - \bar{b}b + g\bar{g} - \bar{g}g.$$

Show that indeed a colour singlet $C = -1$ meson cannot be annihilated through two gluons.

13.5* In problem 12.1 you were invited to obtain the relative three gluon vertices in an SU(2) of colour, and in 12.2 to relate the reduced quark-gluon and three gluon couplings. It is more complicated in an SU(3) of colour, and here it is essential to take account of the (extended) $C$-parity of the gluons. The relative amplitudes may be obtained by creating the colour singlet $C = +1$ combinations of colour and anti-colour between the gluon colours. If you are very energetic, you may do it for all possibilities. I suggest you start with a single gluon, let us say $r\bar{b}$ in a red-blue isospin representation. Show that the $r\bar{b}$ gluon does not couple to the (red-blue) isosinglet gluon and find the relative couplings to

$$\left[\frac{1}{\sqrt{2}}(b\bar{b} - r\bar{r})\right][r\bar{b}], \quad [r\bar{b}]\left[\frac{1}{\sqrt{2}}(b\bar{b} - r\bar{r})\right],$$
$$[r\bar{g}][\bar{b}g], \quad [\bar{b}g][r\bar{g}]$$

(where only the colours have been indicated above).

You could also demonstrate explicitly that the $C = -1$ gluons do not couple to two $C = -1$ gluons (but do couple to the other six in pairs).

[A complete evaluation should yield the results encapsulated in the $8_2 \rightarrow 8 \otimes 8$ table of SU(3) factors found in the Review of Particle Properties published every other year by the Particle Data Group.]

The relation between the quark-gluon ($g$) and three gluon ($G$) reduced couplings can be found by equating the vertex factors for

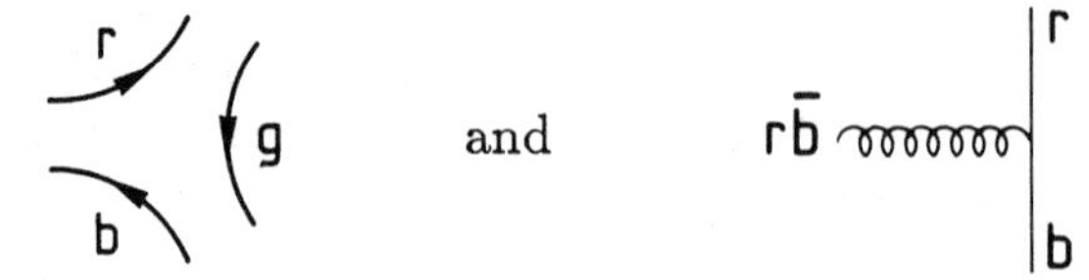


Show that

$$G^2 = \frac{9}{4}g^2$$

This result can also be obtained by considering one gluon loop corrections to the right hand diagram. The sum of such corrections should be equal to $g^2$ times the amplitude for the right hand diagram $\left[\sqrt{\frac{3}{8}}g\right]$ if a hypothetical very soft gluon interacts with a conserved colour current.
[In perturbative QCD the probability that a gluon radiates a gluon is bigger than the probability that a quark radiates a gluon, by a factor 9/4.]

# 14. MASSIVE FERMIONS.

## 14.1 Introduction

The definition of massive fermions, as opposed to light, can be made almost arbitrarily. We choose to call those fermions which were not observed before 1974 massive, and they are the third generation charged lepton $\tau$, the second generation up quark $c$, and the third generation down quark, $b$. There is no doubt now that the third generation up quark, $t$, must be somewhere. It must have a mass in excess of 27 GeV, for it has not been observed in $e^+e^-$ annihilation at the Japanese collider TRISTAN, and the UA1 Collaboration at the CERN $Sp\bar{p}S$ has given a lower limit of 44 GeV.

In this chapter we deal with the physics of the $\tau$ and the mesons containing heavy quarks. We turn first to the $\tau$ lepton.

## 14.2 The $\tau$ lepton

The $\tau$ was discovered after the $1^{--}$ $J/\psi$ $(c\bar{c})$ states and has a mass 1.784 GeV. It has been studied wholly in $e^+e^-$ annihilation (fig.14.1).

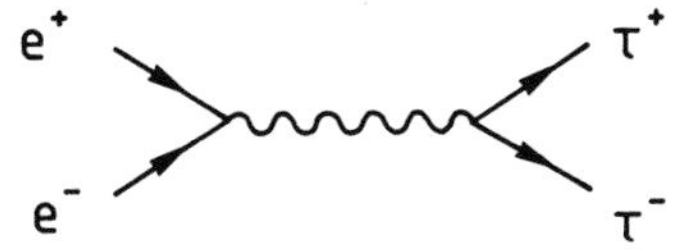


Fig.14.1

The threshold is $\sim$ 3.6 GeV, just below the second $1^{--}(c\bar{c})$ state, $\psi'$. The signature at such low energies was provided by events

$$e^+e^- \rightarrow \mu^+e^-, \mu^-e^+$$

with missing energy. Such events apparently violate lepton conservation, but in fact represent leptonic decay of both $\tau$ leptons. What was actually going on is shown in fig.14.2

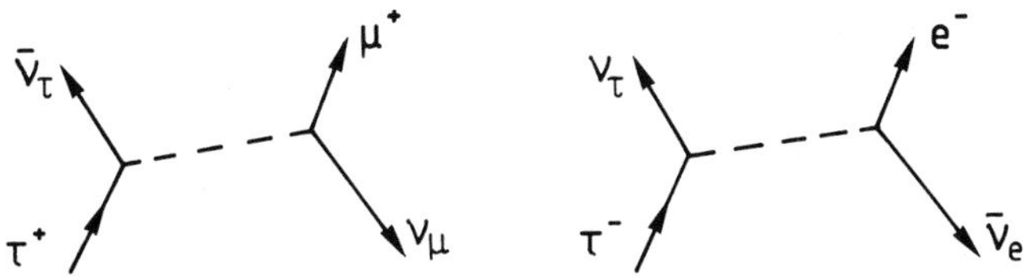


Fig.14.2

The cross section for such processes rises rapidly as the threshold is passed, in agreement with calculations for a lepton pair (we have not calculated this cross section in the low energy limit, but see problem 14.1) and in disagreement with the hypothesis of, say, a pair of bosons. At high energies, $\sim$ 30 GeV, the cross section is found to be the same as for muon pairs and has the $1 + \cos^2\theta$ angular distribution characteristic of spin $1/2$. The $\tau$ is the third generation electron. While the $\mu$ is lighter than any hadron, the $\tau$ has almost twice the mass of the proton and so can (and does) decay into $\nu_\tau + hadrons$ through the process shown in fig.14.3.

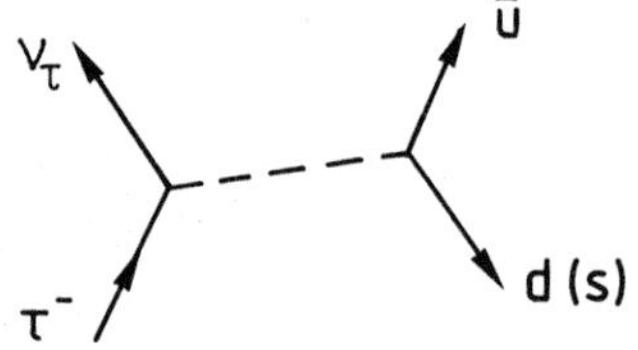


Fig.14.3

The quark pair may be coupled into a stable hadron ($\pi^-$ or $K^-$), a resonance ($\rho$, $a_1$) or simply break up by pair creation in the colour field.

If the weak interaction is universal for the leptons, then the partial decay rates

$$\tau^- \rightarrow \nu_\tau \mu^- \bar{\nu}_\mu, \ \nu_\tau e^- \bar{\nu}_e$$

can be calculated at once from the muon lifetime. The muon decays

$$\mu^- \rightarrow \nu_\mu e^- \bar{\nu}_e$$

and in both $\tau$ decay to leptons and $\mu$ decay to leptons the masses of the final charged leptons can be neglected. Thus we have

$$T_{\tau\rightarrow\mu} = T_{\tau\rightarrow e} = \left(\frac{m_\tau}{m_\mu}\right)^5 T_{\mu\rightarrow e} = \left(\frac{1.784}{0.106}\right)^5 \times \frac{1}{2.2 \times 10^{-6}\text{s}} \tag{14.1}$$

(We made such calculations in Ch.3, but here the treatment is (almost) exact.) This partial decay rate is $6.14 \times 10^{11}\ \text{s}^{-1}$ and the $\tau$ lifetime is given by

$$\frac{1}{\tau_\tau} = \sum_x T_{\tau\rightarrow x} \tag{14.2}$$

and so

$$\tau_\tau = 1.6 \times 10^{-12}\,\text{s} \times \ [\text{Branching ratio } \tau^- \rightarrow \nu_\tau \mu^- \bar{\nu}_\mu] \tag{14.3}$$

The branching ratio can be measured, but we can even make a reasonable estimate of it. To the extent that the quarks coupling to the exchanged $W$ can be treated as plane waves, then with three colours of quark the decays will be distributed in the ratios

$$hadrons : \mu : e = 3 : 1 : 1.$$

Thus we expect (perhaps a bit optimistically) that

$$\frac{\Gamma(\tau^- \rightarrow \nu_\tau \mu^- \bar{\nu}_\mu)}{\Gamma(\tau^- \rightarrow \text{anything})} = \frac{\Gamma(\tau^- \rightarrow \nu_\tau e^- \bar{\nu}_e)}{\Gamma(\tau^- \rightarrow \text{anything})} = 0.2 \tag{14.4}$$

and

$$\tau_\tau \simeq 3 \times 10^{-13}\text{s}.$$

The measured $\mu$ and $e$ branching ratios are both $\sim 0.17-0.18$ and the measured lifetime is $3.0(\pm 0.1)\times 10^{-13}$s.

Lifetime measurements have to be made at high energy, taking advantage of both time dilation and the clean $\tau\bar{\tau}$ signature provided by one $\tau$ decaying into a single charged track and the other into three charged tracks. The decay point of the three can be reconstructed and its distance from the intersection of the colliding $e^+e^-$ beams measured. At a centre of mass energy of 30 GeV the time dilation factor is 8.4 and the mean flight path is 0.76 millimetres. The opening angle of the tracks is only $\sim 10°$ ... Such measurements are hard.

The agreement between the measured $\tau$ lifetime and that predicted using the leptonic branching ratios is satisfactory. It is also possible to predict the branching ratios for certain semi-leptonic modes, in particular $\tau \to \pi\nu_\tau$ and $\tau \to \rho\nu_\tau$. Such calculations (see problem 14.2) use the assumption of universal coupling and the results of other measurements, such as $\pi \to \mu\nu_\mu$. Again the predictions are in good agreement with the data.

It would be very nice to measure accurately $g-2$ for the $\tau$ and see if anything unexpected turned up, but it is not feasible. We think we know everything about the $\tau$—except why it is there, what tauness is, where its mass comes from ...

### 14.3 Hidden charm, hidden bottom

The ONIA states are the heavy quark analogues of positronium; charmonium ($c\bar{c}$) and the unfortunately named bottomonium ($b\bar{b}$). [Perhaps at LEP toponium ($t\bar{t}$) states will appear.] Despite the fact that the first $b\bar{b}$ states were discovered in $pN$ interactions and the $J/\psi$ state which started it all was simultaneously discovered in $pN$ and in $e^+e^-$ annihilation, their study has remained the exclusive province of $e^+e^-$ colliders.

The $J^{PC}=1^{--}$ $s$-wave $Q\bar{Q}$ states couple directly to a virtual photon in $e^+e^-$ annihilation and hence appear as resonances in $e^+e^- \to e^+e^-$, $\mu^+\mu^-$, hadrons (fig.14.4).

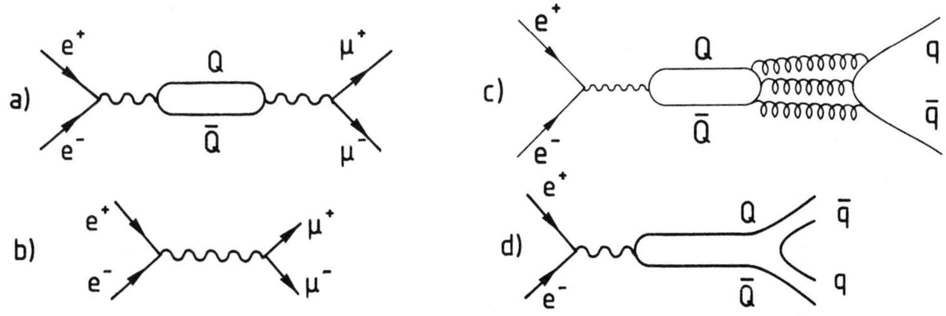


Fig.14.4

Resonant production of $\mu^+\mu^-$ is shown in fig.14.4(a); to this amplitude must be added the non-resonant background amplitude of fig.14.4(b). Resonant production of light hadrons (below threshold for $(Q\bar{q})(\bar{Q}q)$ production) takes place through annihilation into an intermediate state with the quantum numbers of a three gluon colour singlet (fig.14.4(c)). Above threshold the resonance can fall apart into mesons containing one heavy, one light quark. Non-resonant background is of course present in the hadronic final states.

Since the phase of the direct process $e^+e^- \to \mu^+\mu^-$ is constant and the resonance phase varies rapidly through $\pi$, there are marked interference effects which can be readily measured in the clean $\mu^+\mu^-$ state if the energy resolution is sufficient. Such effects are visible in hadron final states as well.

The $c$ quark has a mass $\gtrsim 1.5$ GeV, the $b$ $\gtrsim 5$ GeV. Applying a non-relativistic Schrödinger treatment to bound $Q\bar{Q}$ states is less unjustified than in the case of light quarks $u$, $d$, $s$. The simplest confining potential we could consider is an harmonic potential and in such a potential the states follow a sequence familiar from nuclear physics, fig.14.5

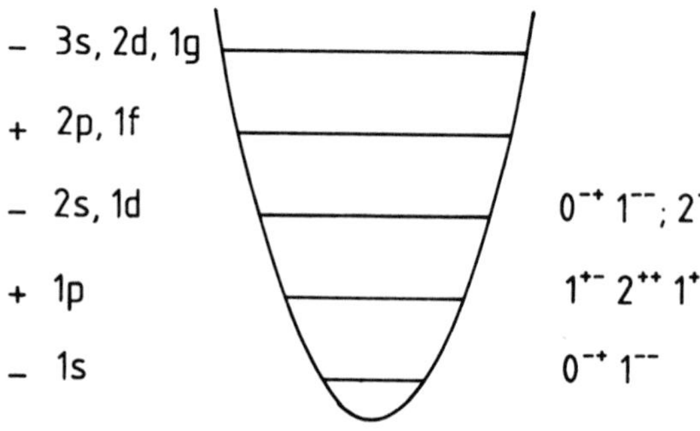


Fig.14.5

The $Q\bar{Q}$ states in a simple harmonic potential are equally spaced and highly degenerate. The $J^{PC}$ composition of the first three levels is shown; the parity of all levels is given on the left of fig.14.5. Remember that the intrinsic parity of a $Q\bar{Q}$ pair is negative. Levels alternate in parity and an $s$-wave $1^{--}$ state occurs every other level.

The potential is NOT harmonic of course, but this scheme is valuable for orientation. In $e^+e^-$ annihilation we expect that the $s$-wave $1^{--}$ states will be excited, having the right $J^{PC}$ for one photon annihilation. The $d$-wave $1^{--}$ states should not be excited, because the wave function vanishes at the origin where $Q$ and $\bar{Q}$ are on top of each other, but the photon couples to $Q\bar{Q}$ at a point. In fact there are spin-dependent forces in QCD (a vector theory) and in the $c\bar{c}$ system they mix quite strongly the 2s, 1d $1^{--}$ states and possibly the 3s, 2d $1^{--}$ states. In the $b\bar{b}$ system such mixing seems to be negligible—remember the $b$ quark is much heavier than the $c$.

In the $b\bar{b}$ system the $\Upsilon(1^{--})$ states have been cleanly identified up to $\Upsilon$4s, which lies only marginally above the threshold for decay into $B\bar{B}$, the $0^{-+}$ states containing a $b$ quark and a $u$ or $d$ quark. The other three lie below this threshold and are very narrow. In the $c\bar{c}$ system, the $\psi(1^{--})$ states have been (probably) identified up to $\psi$4s, but only the first two lie below threshold for $D\bar{D}$ and there is certainly some $s-d$ mixing taking place so that the higher states are not so cleanly identifiable. The known $1^{--}$ states are listed in Table 14.1

**Table 14.1**

| Name | State | Mass MeV | Width MeV | Partial Width $e^+e^-$ or $\mu^+\mu^-$ KeV |
|---|---|---|---|---|
| | | $c\bar{c}$ $\psi$ $1^{--}$ states | | |
| $J/\psi$ | 1s | 3097 | 0.063 | 4.7 |
| $\psi'$ | 2s,1d | 3686 | 0.215 | 2.1 |
| $\psi''$ | | 3770 | 25 | 0.26 |
| $\psi$ | 3s,2d (?) | 4030 | 52 | 0.75 |
| $\psi$ | | 4159 | 78 | 0.77 |
| $\psi$ | ~4s | 4415 | 43 | 0.47 |
| | | $b\bar{b}$ $\Upsilon$ $1^{--}$ states | | |
| $\Upsilon$ | 1s | 9460 | 0.043 | 1.22 |
| | 2s | 10023 | 0.030 | 0.54 |
| | 3s | 10355 | ~0.012 | 0.40 |
| | 4s | 10577 | 24 | 0.24 |
| | [ 5s | 10865 | 110 | ~0.3 ] |
| | [ 6s | 11019 | ~80 | ~0.1 ] |

Note that there is enough information in the table above to work out the masses of $\psi$2s, $\psi$1d and the mixing angle between them. Note also that we have reached as far as the third radial excitation of the $\psi$1s, $\Upsilon$1s states. In the light quark sector no radial excitations have been unambiguously identified—in any channel there are many broad states overlapping. The ONIA states provide in comparison a clean laboratory for testing our ideas about hadron structure. There is another very nice feature. Between the $\psi'$ (~2s) at 3.686 GeV and the $\psi$(1s) at 3.097 GeV, four positive parity states are expected, $1^{+-}$; $2^{++}$, $1^{++}$, $0^{++}$. The three with positive $C$ can be reached by $E1$ electromagnetic transitions from the (narrow) $\psi'$

$$\psi' \to \chi\gamma$$

[The $1^{+-}$ state cannot be reached in this way because it has the same $C$ as $\psi'$ and the photon is odd under $C$.] The states are there and have been seen. In $b\bar{b}$, there are again such $p$-wave states between the $s$-states. Both $\Upsilon$2s AND $\Upsilon$3s are narrow and

$$\Upsilon 3\text{s} \to \chi_b(2p)\gamma$$

was observed before

$$\Upsilon 2\text{s} \to \chi_b(1p)\gamma$$

(an accident of machine scheduling).

The spectroscopy of charmonium and bottomonium is displayed in fig.14.6. Note that in addition to the triplet-triplet $E1$ electromagnetic transitions, $M1$ transitions to the lowest charmonium state $\eta_c$ $0^{-+}$ have been observed, and

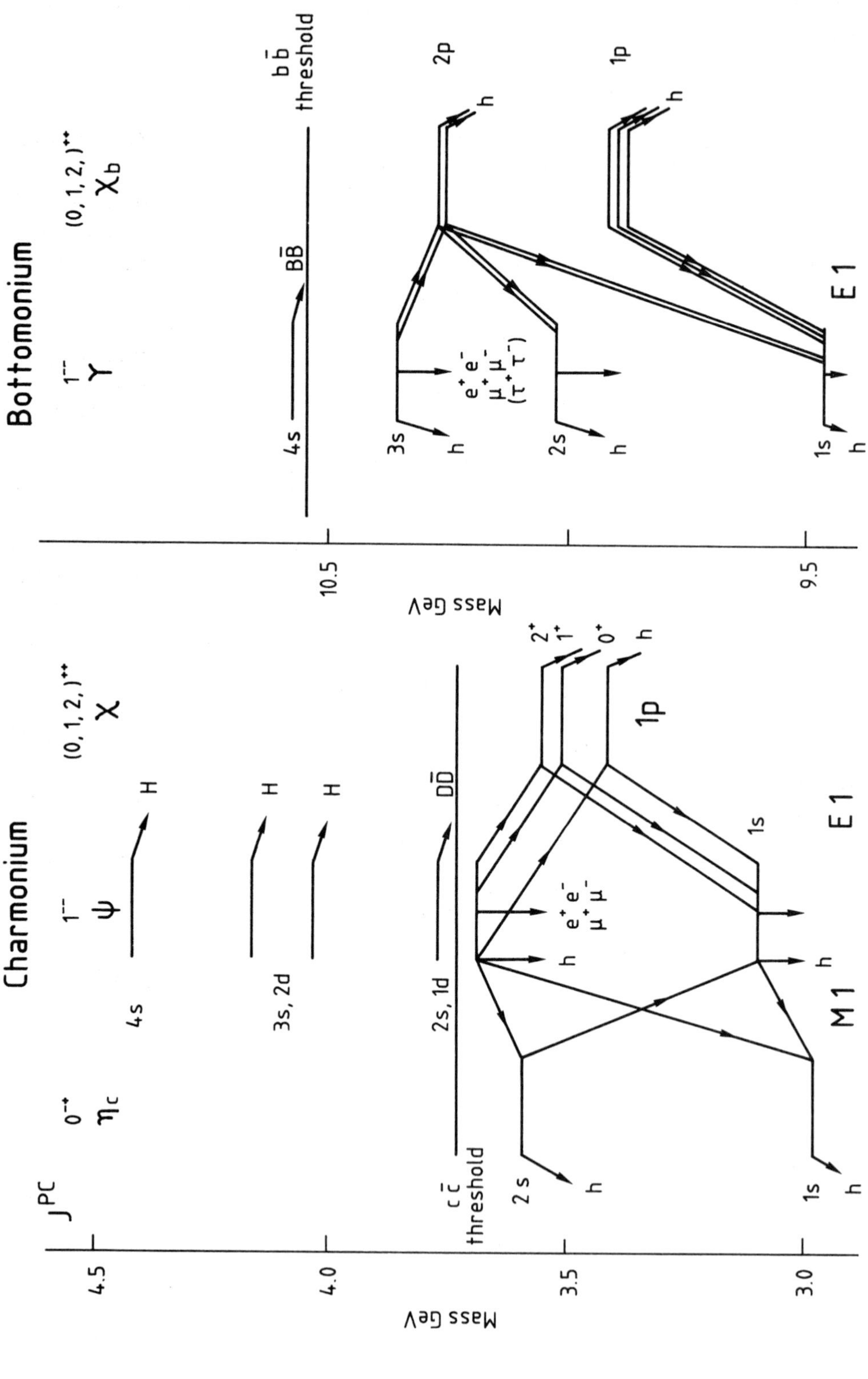

Charmonium
Bottomonium
$J^{PC}$
$0^{-+}$
$\eta_c$
$1^{--}$
$\psi$
$(0,1,2,)^{++}$
$\chi$
$\Upsilon$
$\chi_b$
4s
3s, 2d
2s, 1d
2 s
1s
1p
2p
3s
2s
H
h
$D\bar{D}$
$B\bar{B}$
$c\bar{c}$ threshold
$b\bar{b}$ threshold
$e^+e^-$
$\mu^+\mu^-$
$(\tau^+\tau^-)$
$2^+$
$1^+$
$0^+$
M1
E1
Mass GeV
4.5
4.0
3.5
3.0
10.5
9.5


Fig.14.6

(probably) to the first radial excitation $\eta_c'$. The corresponding $b\bar{b}$ states have not so far been identified.

### 14.4 The ONIA potential

It is clear from fig.14.6 that the effective potential between $Q$ and $\bar{Q}$ has some of the features of an harmonic potential, but the agreement is far from perfect. The 1p states lie closer to the 2s states than to the 1s states and successive $s$ states are not equally spaced. The spacing of equivalent pair of states has little dependence on the heavy quark mass. These observations already constrain the form of the potential significantly.

The amount of information contained in these simple systems is huge. First, the centres of the groups of levels corresponding to each principal quantum number can be fitted with a plausible potential—one which works very well is the Cornell potential [E. Eichten *et al*, *Phys. Rev.* **D17** 3090 (1979); **D21** 203 (1980)]

$$V = -\frac{\alpha_s}{r} + \kappa r \tag{14.5}$$

where the first term is the short range colour coulomb potential and the second is the flux tube energy. The colour coupling $\alpha_s$ weakens as $r$ diminishes in accord with the prescription of QCD (the field carries colour and so smears it out around the quark.) The fitted value of $\kappa$ is 0.18 GeV$^2$—just the value obtained from the light quark Regge trajectories. Secondly, the $Q\bar{Q}$ wave function can be calculated from the potential, and provided with that information the partial decay rates $\psi$, $\Upsilon \to e^+e^-$, $\psi \to \chi\gamma$, $\Upsilon \to \chi_b\gamma$, $\chi \to \psi\gamma$, $\chi_b \to \Upsilon\gamma$ can be calculated and compared with measurements. It all works very well—after relativistic corrections have been made in the $c\bar{c}$ system. We shall have to content ourselves with a few simple calculations.

Suppose

$$E = 2m - \frac{\alpha_s}{r} + \kappa r + \frac{p^2}{m} \tag{14.6}$$

and $pr = n(\hbar)$. For each (integer) $n$ minimise the energy, calculate the radius and hence the mass of the $c\bar{c}$, $b\bar{b}$ system. (This recipe gives the correct level spacing for coulomb and harmonic oscillator potentials.)

Then

$$E = 2m - \frac{\alpha_s}{r} + \kappa r + \frac{n^2}{mr^2} \tag{14.7}$$

and setting

$$\frac{\partial E}{\partial r} = 0$$

yields

$$\frac{2n^2}{mr_n^3} = \frac{\alpha_s}{r_n^2} + \kappa \tag{14.8}$$

which is easily solved for $r_n$, given $\kappa$, $\alpha_s$, $m$. The energy takes the form

$$E = \frac{3}{2}\kappa r_n - \frac{1}{2}\frac{\alpha_s}{r_n} + 2m. \tag{14.9}$$

I have taken $\alpha_s = 0.5$, $\kappa = 0.18\ \text{GeV}^2$, $m_c = 1.5$ GeV, $m_b = 5$ GeV and obtained

**Table 14.2**

| $n$ | $r_n\ (c\bar{c})$ | $r_n\ (b\bar{b})$ GeV$^{-1}$ |
|---|---|---|
| 1 | 1.5 (0.3 fm) | 0.7 |
| 2 | 2.8 | 1.65 |
| 3 | 3.85 | 2.3 |
| 4 | 4.7 | 3.0 |
| 5 | 5.55 | 3.6 |
| 7 | 7 | 4.6 |

[Divide by ~ 5 to obtain the radii in fm.] The masses are then redetermined from the energy of the ground state and that fixes the whole sequence. The results I calculated ($m_c \to 1.48$ GeV, $m_b \to 4.81$ GeV) are given in Table 14.3

**Table 14.3**

| | | $M(c\bar{c})$ | | | $M(b\bar{b})$ | |
|---|---|---|---|---|---|---|
| n | calc. | state | Exp | calc. | state | Exp |
| 1 | <u>3.05</u> | 1s | 3.05 | <u>9.45</u> | 1s | 9.46 |
| 2 | 3.48 | 1p$\chi$ | ~3.48 | 9.91 | 1p | ~ 9.9 |
| 3 | 3.78 | 2s,1d | ~3.7 | 10.13 | 2s | 10.03 |
| 4 | 4.02 | 2p | – | 10.35 | 2p | 10.25 |
| 5 | 4.26 | 3s,2d | ~4.1 | 10.52 | 3s | 10.35 |
| 7 | 4.66 | 4s | 4.4 | 10.80 | 4s | 10.58 |

The agreement between calculation and experiment is quite good for the first few states, but becomes progressively worse as $n$ increases. The discrepancy is, for both $c\bar{c}$ and $b\bar{b}$, about 200 MeV for $n = 7$. The most important features of the comparison are that the spacings are about right at both ends of the scale and the description is equally good (or bad) for both cases.

We could not expect such a simple calculation to do better, and in fact fitting $\alpha_s$ and the quark masses simultaneously to all levels does not produce a significantly better result. You should note that the assignment of states to a particular value of $n$ in Table 14.3 has been made according to an harmonic oscillator scheme and the same degeneracies do not hold with the potential (14.5). You may investigate other forms for the potential using the simple methods of (14.6–9) [Problem 14.4] but it is necessary to solve the Schrödinger equation for a given form in order to obtain accurate energies, and the corresponding wavefunctions, which are needed in order to calculate chromomagnetic splittings and electromagnetic decay rates. We can only make rough estimates using the results of Table 14.2.

### 14.5 Chromomagnetic splitting in heavy quark systems

The states $\psi$ and $\eta_c$ are $s$-waves, spin triplet and singlet respectively. The splitting is chromomagnetic and should scale from the 630 MeV $\pi - \rho$ splitting:

$$M_\psi - M_{\eta_c} = 630\,\text{MeV}\left(\frac{m}{m_c}\right)^2 \frac{|\psi_\psi(0)|^2}{|\psi_\rho(0)|^2} \tag{14.10}$$

We take the constituent light quark mass, $m \sim 300$ MeV. We also have

$$|\psi(0)|^2 \propto \frac{1}{r^3}$$

where $r$ is the Bohr radius of the state. The splitting between $\psi$ and $\eta_c$ is thus expected to be

$$630 \times \left(\frac{1}{5}\right)^2 \times \left(\frac{0.5\,\text{fm}}{0.3\,\text{fm}}\right)^3 = 116 \text{ MeV}.$$

assuming a radius $\sim$ 0.5fm for the light quark 1s system. The measured splitting is 120 MeV

Chromomagnetic splitting is also observed in $Q\bar{q}$ systems and again a very simple calculation gives results in good agreement with the data. The lowest lying $Q\bar{q}$ mesons are the 1s states:

| | $c\bar{d}$ | $c\bar{u}$ | $c\bar{s}$ |
|---|---|---|---|
| $0^-$ | $D^+(1869)$ | $D^\circ(1865)$ | $D_s^+(1971)$ |
| $1^-$ | $D^{*+}(2010)$ | $D^{*\circ}(2007)$ | |

| | $b\bar{u}$ | $b\bar{d}$ | $b\bar{s}$ |
|---|---|---|---|
| $0^-$ | $B^-(5271)$ | $B^\circ(5275)$ | |
| $1^-$ | $B^{*-}(\sim 5325)$ | $B^{*\circ}(\sim 5325)$ | |

The states $Q\bar{u}$, $Q\bar{d}$ form doublets of $(u-d)$ isospin; states $Q\bar{s}$ are isospin singlets. The $0^-$ states can only decay weakly, but

$$D^* \to D\pi, D\gamma$$
$$B^* \to B\gamma$$

The chromomagnetic splittings are

$$D^* - D = 143 \text{ MeV}$$
$$B^* - B = \ \ 52 \text{ MeV}$$

We would expect the splittings to be

$$\sim 630 \text{ MeV} \times \left(\frac{m}{m_c}, \frac{m}{m_b}\right) = 126,\ 38 \text{ MeV}$$

and this expectation is in very good agreement with the measurements, considering the crude calculation and the sensitivity to the wave function at the origin. [The $D^* - D$ and $B^* - B$ splittings were calculated above assuming the same value for the wave function at the origin as for the light quark 1s states and so are expected to be low.]

### 14.6 Leptonic decays of 1s $\psi$, $\Upsilon$

We will conclude with one other simple (and approximate) calculation: decay of $\psi$, $\Upsilon$ into $e^+e^-$, $\mu^+\mu^-$.

The partial width is given by the Golden Rule

$$\Gamma_{\ell^+\ell^-} = 2\pi|\mathcal{M}|^2\rho$$

and the matrix element $\mathcal{M}$ is to be calculated from the diagram fig.14.7

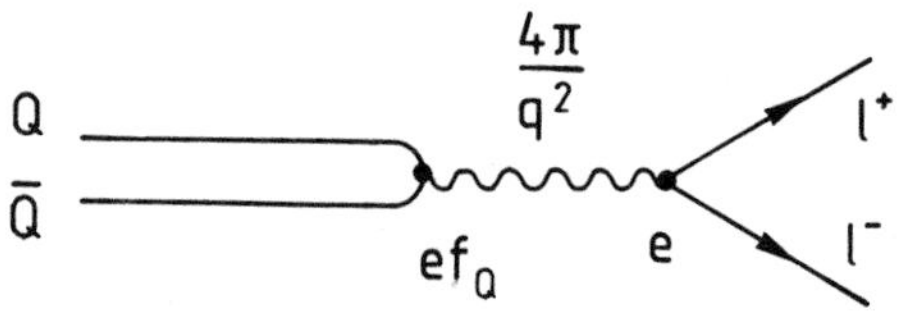


Fig.14.7

Working with the plane waves normalised to 1 per unit volume, in the centre of mass

$$\rho \simeq \frac{4\pi p^2}{(2\pi)^3}\frac{dp}{dE}$$

$$\mathcal{M} \sim \frac{4\pi e^2 f_q}{q^2}\psi(0) \qquad |q^2| = M_V^2 \tag{14.11}$$

whence

$$\Gamma_{\ell^+\ell^-} \sim 2\pi\left(\frac{4\pi}{2\pi}\right)^3 \frac{e^4 f_q^2}{M_V^4} p^2 \frac{dp}{dE}|\psi(0)|^2 \tag{14.12}$$

The leptons are relativistic, so $dp/dE = \frac{1}{2}$, $p = \frac{M_V}{2}$

$$\Gamma_{\ell^+\ell^-} \sim 2\pi e^4 f_q^2 \frac{|\psi(0)|^2}{M_V^2} \tag{14.13}$$

Note dimensions — $|\psi(0)|^2$ is 1/volume and has dimension GeV$^3$. Then

$$\Gamma_{\ell^+\ell^-} \sim \frac{2\pi e^4 f_q^2}{r^3 M_V^2} \tag{14.14}$$

where $r^3$ is a measure of the volume of the bound state (in GeV$^{-3}$). Simply inserting the radii from Table 14.2 we find for $J/\psi$

$$\Gamma_{\ell^+\ell^-} = 2\pi\left(\frac{1}{137}\right)^2 \frac{4}{9}\frac{1}{(1.5)^3(3.1)^2} = 4.6\times 10^{-6}\ \text{GeV}$$

$$= 4.6\ \text{KeV (Exp: 4.7 KeV)}$$

and for $\Upsilon 1s$

$$2\pi\left(\frac{1}{137}\right)^2\frac{1}{9}\frac{1}{(9.46)^2(0.7)^3} = 1.2\times10^{-6}\ \text{GeV (Exp : 1.22 KeV)}$$

The excellent agreement is fortuitous, since the right hand side of (14.13) is actually too small by a factor of eight. [A factor three comes from colour and a factor 8/3 from evaluating the spinor products in the matrix element properly; problem 14.5.] The replacement of $|\psi(0)|^2$ by $r^{-3}$ in order to obtain (14.14) is merely order of magnitude correct and dimensionally correct. It is however significant that the calculation yields the ratio of leptonic widths of $\Upsilon 1s$ to $J/\psi$ correctly.

The spectroscopy of the $c\bar{c}$ and $b\bar{b}$ systems seems to be entirely in accord with our ideas of what QCD should yield for these systems of massive quarks. The colour fields seem to be universal, indifferent to the flavour of the source.

### 14.7 References

Our knowledge of the physics of massive fermions has been developing rapidly and review articles become out of date very quickly. With that warning, we give below some readily accessible articles.

The tau lepton, M.L. Perl, *Ann. Rev. Nucl. Part. Sci.* **30** 299 (1980)

Psionic matter, W. Chinowsky, *Ann. Rev. Nucl. Part. Sci.* **27** 393 (1977)

Charm and beyond, T. Appelquist, R.M. Barnett, K. Lane, *Ann. Rev. Nucl. Part. Sci.* **28** 387 (1978)

Upsilon resonances, P. Franzini, J. Lee-Franzini, *Ann. Rev. Nucl. Part. Sci.* **33** 1 (1983)

Upsilon spectroscopy at CESR, K. Berkelman, *Phys. Rep.* **98** 145 (1983)

Heavy quark systems, W. Kwong, J. Rosner, C. Quigg, *Ann. Rev. Nucl. Part. Sci.* **37** 325 (1987)

## Problems

14.1* The rapid rise of the cross section for

$$e^+e^- \to \tau^+\tau^-$$

as the threshold is exceeded was an important indication that the spin of the $\tau$ is $\frac{1}{2}$. In Ch.9 the cross section was calculated for energies far above threshold where the helicity states become states of definite handedness. It is easy to see that for massive outgoing fermions the helicity combinations $\bar{L}_h R_h$ and $\bar{R}_h L_h$ give the same contribution to the matrix element squared, summed and averaged, as the handed combinations $\bar{L}R$ and $\bar{R}L$, namely

$$2\left(\cos^4\frac{\theta}{2}+\sin^4\frac{\theta}{2}\right)$$

It is also fairly obvious that the combinations $\bar{L}_h L_h$ and $\bar{R}_h R_h$ will be proportional to

$$\frac{K^2}{(1+K^2)^2}\sin^2\frac{\theta}{2}\cos^2\frac{\theta}{2}$$

where $K^2 = (1-\beta)/(1+\beta)$. The coefficient of the above term is not quite obvious. Show that when the mass of the outgoing lepton cannot be ignored, the effect is

$$1+\cos^2\theta \to 2-\beta^2+\beta^2\cos^2\theta$$

and hence that

$$\sigma = \frac{4\pi}{3}\frac{e^4}{s}\frac{\beta(3-\beta^2)}{2}$$

where $\beta$ is the velocity of the massive outgoing leptons.

Consider the form that the electromagnetic current for scalar particles must take, and infer the $\beta$ dependence for production of a pair of scalar particles. You could even check the angular dependence.

[This problem is not very difficult, but requires careful book keeping and a clear head. If the $z$ axis is taken along the $e^+e^-$ beams, only $\alpha_x$ and $\alpha_y$ contribute. Remember that the large and small (Weyl) components behave in the same way under rotations. It takes about as long to solve this problem directly using the methods of Ch.9 as it takes to do it with the usual apparatus of projection operators and trace techniques in $\gamma$ matrix algebra, but these current-current (lowest order) interactions are simple.]

14.2* If the $(\tau, \nu_\tau)$ pair represent another family of leptons with the universal weak couplings, then we can predict not only the partial leptonic decay rates but at least some of semi-leptonic rates. For example, the partial decay rate for $\tau \to \pi\nu_\tau$ can be calculated from the measured rate for $\pi \to \mu\nu_\mu$. In both cases the leptonic part has the form

$$\sqrt{2}G\bar{u}\gamma_\mu\frac{(1-\gamma_5)}{2}u$$

and this must be contracted with a four-vector representing the pion. The matrix element for both processes therefore takes the form

$$\sqrt{2}Gf_\pi\bar{u}p_\mu\gamma_\mu\frac{(1-\gamma_5)}{2}u$$

(the above expression defines the quantity $f_\pi$).

It is very quick and simple to evaluate the matrix element for pion decay, working in the centre of mass of the pion. With $\sqrt{2E}$ normalisation the result is

$$\mathcal{M}^2_{\pi\to\mu\nu} = 2G^2 f_\pi^2 m_\mu^2(m_\pi^2 - m_\mu^2)$$

This can be obtained using the helicity states given in Ch.8, without any difficulty. The $(1-\gamma_5)$ piece reduces the problem to one of multiplying $2\times 2$ matrices. Do it.

It is a little trickier to evaluate the matrix element for $\tau \rightarrow \pi\nu_\tau$, if we use the methods of Ch.8 and do not adopt a manifestly covariant formalism. If you use the helicity states of Ch.8, it is important to remember that (i) they are set up such that the $z$ axis is along the lepton direction (ii) that decay can take place over the whole angular range for the $\tau$ spin relative to the outgoing particles. If you are careful about this, then working in the $\tau$ centre of mass it becomes easy to show that

$$\begin{aligned}\overline{\mathcal{M}^2_{\tau\rightarrow\pi\nu}} &= 2G^2 f_\pi^2 E_\nu m_\tau (E_\pi + p_\pi)^2 \\ &= G^2 f_\pi^2 m_\tau^2 (m_\tau^2 - m_\pi^2)\end{aligned}$$

after averaging over all orientations. Do it.

Assuming that $f_\pi^2$ is a constant, independent of momentum transfer to the pion, show that

$$\Gamma_{\pi\rightarrow\mu\nu} = \frac{G^2 f_\pi^2}{8\pi} m_\pi m_\mu^2 \left(1 - \frac{m_\mu^2}{m_\pi^2}\right)^2$$

$$\Gamma_{\tau\rightarrow\pi\nu} = \frac{G^2 f_\pi^2}{16\pi} m_\tau^3 \left(1 - \frac{m_\pi^2}{m_\tau^2}\right)^2$$

[Remember the lepton normalisation.] Hence calculate $\Gamma_{\tau\rightarrow\pi\nu}$ and estimate the branching ratio.

Consider what could be done to predict the branching ratio for $\tau^- \rightarrow \rho^-\nu_\tau$. [Remember $\rho^\circ \rightarrow e^+e^-$ and the magic words Conserved Vector Current.]

14.3 Here are three fairly easy exercises.

(i) Estimate the chromomagnetic splitting between the ground state spin singlet and spin triplet $b\bar{b}$ states.

(ii) The $\psi(3685)$ and $\psi(3770)$ are separated in mass by only 84 MeV. Both are excited in $e^+e^-$ annihilation, and must be mixtures of the second $1^-s$ and first $1^-d$ $c\bar{c}$ states. The partial widths into $e^+e^-$ are 2 keV and (approximately) 0.25 keV respectively.

Write these two states as orthogonal mixtures of the $1^-s$ and $1^-d$ states. Find the composition of $\psi(3685)$ and $\psi(3770)$ (the mixing angle) and find the unmixed masses of $1^-s$ and $1^-d$. [Assume $1^-d$ does not couple to $e^+e^-$ because of the centrifugal barrier.]

(iii) Calculate the cross section for $e^+e^- \rightarrow \mu^+\mu^-$ in the vicinity of the $J/\psi$ resonance. [Remember the interference between the resonant term and single photon exchange background.]

14.4 This is an invitation to investigate in more detail the curious aspects of the effective potential in ONIA. Consider the masses given in Tables 14.1, 14.3 and the display in fig.14.6. First, it should be trivial to demonstrate that a coulomb potential is totally unsatisfactory. Then repeat the program of section 14.4 for a simple harmonic oscillator potential, finding the best value for the spring constant you can. Show that for a common spring constant the spacing of levels of given $n$ is proportional to $m_Q^{-\frac{1}{2}}$. Finally study a logarithmic potential. Show that the level spacing is independent of $m_Q$. Compare your best logarithmic potential with the Cornell form given, over the range probed by ONIA, $\sim 0.1 - 1$ fm. [Plot the two forms.]

14.5 In section 14.6 we made a rough estimate of the width for decay of a vector meson into a charged lepton pair, assuming non-relativistic quarks. The details of the electromagnetic matrix element were ignored. If we take the $m = \pm 1$ vector meson spin states, then a non-relativistic pair annihilating into $\ell^+\ell^-$ leads to our old friend

$$2(1 + \cos^2\theta)$$

multiplying the square of the approximate matrix element (14.11). This result is almost obvious, but verify it and hence show that for one flavour of quark and three colours

$$\Gamma_{\ell^+\ell^-} = 16\pi e^4 f_q^2 \frac{|\psi(0)|^2}{M_V^2}$$

(This is often referred to as the Van Royen-Weisskopf formula).

## 15. WEAK INTERACTIONS OF QUARKS.

### 15.1 The pattern of quark decay

The weak interactions of quarks exhibit a curious but simple pattern. The charge changing weak interactions turn $u$ into $d$, $c$ into $s$ and (undoubtedly) $t$ into $b$. If this were all, the pattern would be both simple and elegant and would mirror the lepton sector. However, the lightest strange and bottom hadrons would be stable and the world would be very different. This simple pattern is however almost realised: the weak interactions couple, with universal strength, $u$ to $d'$, $c$ to $s'$ and $t$ to $b'$, where $d'$, $s'$ and $b'$ are orthogonal linear mixtures of $d$, $s$, $b$ and predominantly $d$, $s$, $b$ respectively. There is another remarkable regularity. The weak neutral current interactions do not change lepton or quark charge; they do not change flavour either.

### 15.2 The Cabibbo angle

The diagrams for the decay of the muon and for nuclear $\beta$ decay are respectively figs. 15.1(a) and (b).

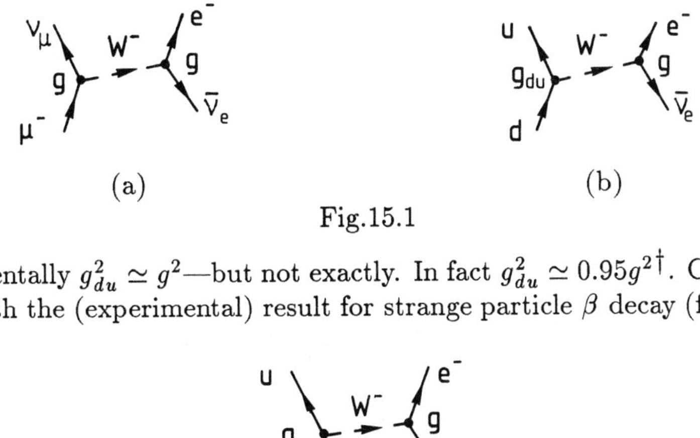


Fig.15.1

Experimentally $g_{du}^2 \simeq g^2$—but not exactly. In fact $g_{du}^2 \simeq 0.95g^2$†. Compare this result with the (experimental) result for strange particle $\beta$ decay (fig.15.2)

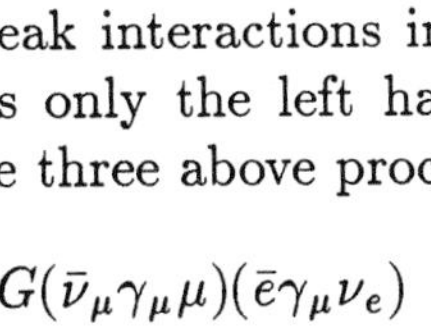


Fig.15.2

where $g_{su}^2 \simeq 0.05g^2$. These weak interactions involve spin 1 $W^\pm$ coupling to a weak current which contains only the left handed projections of fermions. Ignoring the mass of the $W$ the three above processes have matrix elements

$$G(\bar{\nu}_\mu \gamma_\mu \mu)(\bar{e}\gamma_\mu \nu_e)$$
$$G_{du}(\bar{u}\gamma_\mu d)(\bar{e}\gamma_\mu \nu_e)$$
$$G_{su}(\bar{u}\gamma_\mu s)(\bar{e}\gamma_\mu \nu_e)$$

where it is to be understood that only the left handed projections of fermions take part—this saves having to write lots of factors of $(1-\gamma_5)$.

† $g_{du}^2$ is obtained from superallowed Fermi transitions, such as ${}^{14}\mathrm{O} \to {}^{14}\mathrm{N}^* e^+ \nu_e$.

Suppose the weak interaction couples $W^{\pm}$ to a new isospin doublet $\binom{u}{d'}$ and the strength is the universal $G$. Then $d'$ must be a mixture of $d$ and $s$ and the most elegant choice is

$$d' = d\cos\theta_c + s\sin\theta_c \tag{15.1}$$

with coupling

$$G(\bar{u}\gamma_\mu d')$$

Then

$$G_{du} = G\cos\theta_c, \; G_{su} = G\sin\theta_c$$

and with

$$\sin^2\theta_c = 0.05$$

the $d \to u$ and $s \to u$ transitions saturate the weak coupling. The mixing angle $\theta_c$ is called the Cabibbo angle. We have no idea of what it really means physically.

There is of course an orthogonal mixture

$$s' = -d\sin\theta_c + s\cos\theta_c \tag{15.2}$$

It would be very nice if this formed a weak isospin doublet with the $c$ quark. Then in addition to a weak quark current

$$G(\bar{u}\gamma_\mu d')$$

we would have another

$$G(\bar{c}\gamma_\mu s')$$

and this time $G^2_{cs} = 0.95G^2$ while $G^2_{cd} = 0.05G^2$.

### 15.3 Charm decay

A remarkable pattern should appear in the weak decays of $c$ quarks. The lowest mass hadrons which carry a $c$ quark are

$$0^- : D^+(c\bar{d}) \text{ mass } 1.869 \text{ GeV}$$

and

$$D^\circ(c\bar{u}) \text{ mass } 1.865 \text{ GeV.}$$

These can only decay by turning the $c$ quark into a lighter, and $\sim 95\%$ of the time it should be $c \to s$ where the $W$ couples to $\mu^+\nu_\mu$, $e^+\nu_e$, $ud'$ (fig.15.3)

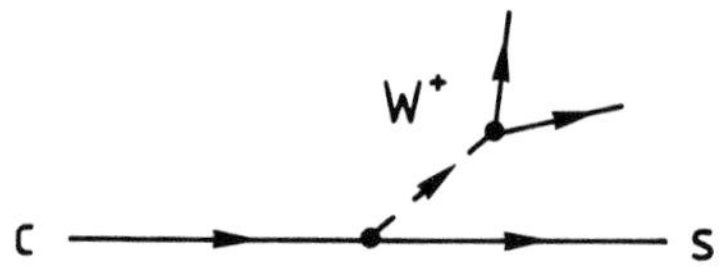


Fig.15.3

For semi leptonic decays

$$D \to e^+\nu_e X,\ D \to \mu^+\nu_\mu X$$

the energy released is $\lesssim 1.3$ GeV (because of the strange quark). These decay rates can be calculated from the muon decay rate, and a crude estimate is

$$T_{D\to\ell} \simeq \left(\frac{1.3\ \text{GeV}}{m_\mu}\right)^5 \frac{1}{\tau_\mu} = \left(\frac{1.3}{0.106}\right)^5 \frac{1}{2\times 10^{-6}} = 1.4\times 10^{11}\ \text{s}^{-1} \qquad (15.3)$$

Since the $W$ couples to $\mu$, $e$ and three colours of $ud'$ this should be $\sim 1/5$ of the total decay rate. Then we estimate

$$T_{D\to x} = 7\times 10^{11}\ \text{s}^{-1} \quad ; \quad \tau_D = 1.4\times 10^{-12}\ \text{s}$$

Experimentally $\tau_{D^+} \simeq 1.09\times 10^{-12}$ s and the branching ratio $D^+ \to e^+X$ is $\sim 19\%$, in remarkable agreement with the simple calculation. However

$$\tau_{D^\circ} \simeq 0.42\times 10^{-12}\ \text{s} \qquad BR \quad D^\circ \to e^+X \sim 5\%$$

The branching ratios for $c \to e^+X$, $\mu^+X$ are both $\sim 8\%$, averaged over all weakly decaying charmed particles produced at high energy ($D^\circ$, $D^+$, $D_s^+$, $\Lambda_c^+$ ...). What is clear is that the hadronic decays are enhanced over the simple spectator model of fig.15.4

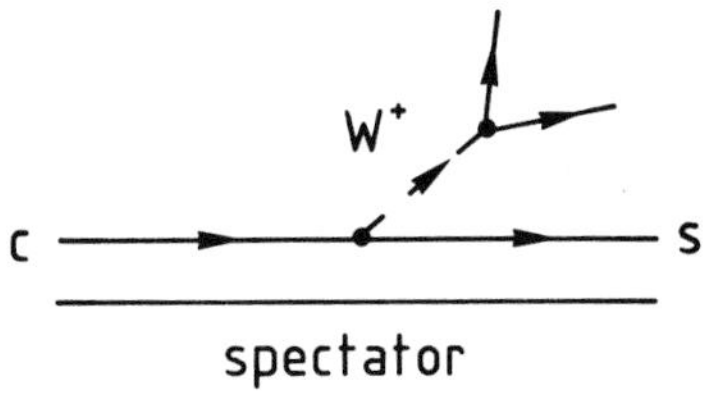

Fig.15.4

particularly in $D^\circ$ decay. There is an additional weak decay mechanism for $D^\circ$ decay, which does not exist for $D^+$, fig.15.5

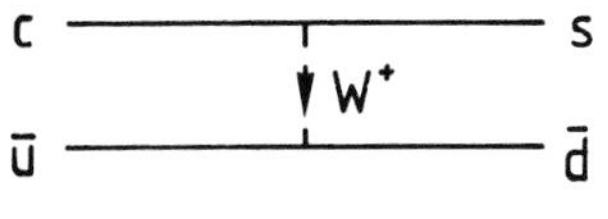

Fig.15.5

(where the weak interaction is followed by quark pair creation.) It is not obvious that such diagrams will be important, but there is evidence that such a mechanism does operate in hadronic decays of $D^\circ$. The branching ratio for $D^\circ \to K^\circ\phi$ is $\sim 1.4\%$ and the only feasible mechanism is that of fig.15.6

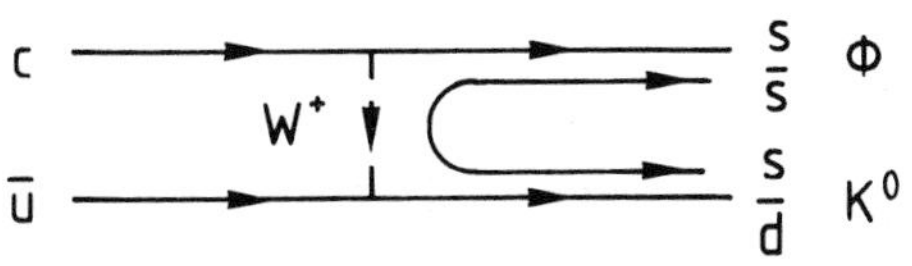

Fig.15.6

There is also a $0^-$ $c\bar{s}$ meson — a charged isosinglet $D_s^+$ with a mean lifetime of $0.47 \times 10^{-12}$ s. This decays through the diagram of fig.15.7

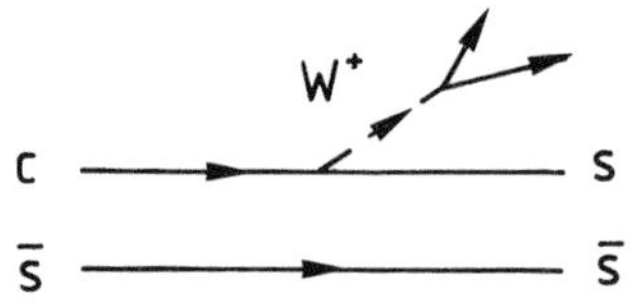


Fig.15.7

and has been identified in $\phi\pi$. Other possible final states are $\eta$, $\eta'$ + pions — $\eta$ and $\eta'$ are rich in $s\bar{s}$ pairs.

A diagram which at first sight could be important for $D_s^+$ decay is

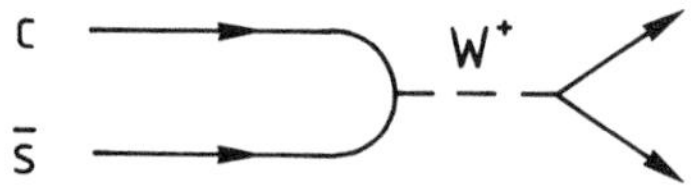


Fig.15.8

This is not important, because for a $0^-$ state fermion and anti-fermion have the same helicities—and $\gamma_\mu$ couples opposite handed states. This mode (and the corresponding Cabibbo suppressed decay for $D^+$) are negligible for the same reason that $\pi \to e\nu/\pi \to \mu\nu \sim 10^{-4}$.

**15.4 Bottom decay**

The $B$ mesons, spin singlet $0^-$ combinations of the $b$ quark, have been identified:

$$B^-(b\bar{u}) \quad \text{mass } 5.271 \text{ GeV}$$

$$B^\circ(b\bar{d}) \quad \text{mass } 5.275 \text{ GeV}$$

but their individual lifetimes have not been measured. The average lifetime of all weakly decaying hadrons which carry a $b$ quark has been measured, and the averaged branching ratios for $b \to e^-X$, $b \to \mu^-X$ have been measured and are 12%. The lifetime is $\sim 1.2 \times 10^{-12}$ s. These mesons are the lightest which carry a $b$ quark, and can only decay by changing the $b$ into a lighter quark $c$ (in the second generation) or $u$ (in the first). The transition $b \to c$ is dominant. We know this because $B$ decay final states are rich in kaons and also from the high energy tails of the lepton spectra in $B \to e + X$, $\mu + X$. Since the $B$ has mass 5.272 GeV, the maximum energy released in $b \to c$ is $\sim 3.4$ GeV, $b \to u \approx 5$ GeV.

Then we expect

$$T_{B\to\ell} \simeq \left(\frac{3.4}{m_\mu}\right)^5 \frac{1}{\tau_\mu}\frac{G_{bc}^2}{G^2} = 1.7 \times 10^{13}\frac{G_{bc}^2}{G^2} \tag{15.4}$$

With a 12% branching ratio to each of $e$, $\mu$

$$T_{B\to\ell} \simeq \left[\frac{1}{\tau_b} = 0.83 \times 10^{12}\right] \times 0.12 = 10^{11}\ \text{s}^{-1}$$

so that

$$\frac{G_{bc}^2}{G^2} \simeq \frac{1}{200} \qquad (\text{compare } \frac{G_{su}^2}{G^2} \simeq \frac{1}{20})$$

The $(e, \mu)$ leptonic branching ratios are expected to be 17% in the spectator model, a little less than the 20% calculated for $W$ coupling to $e\nu$, $\mu\nu$, $ud'$ final states because the more massive $\tau\nu$, $cs'$ make a small contribution (phase space suppressed). Again the hadronic decay modes are probably enhanced somewhat over the simple spectator prediction and for $B^\circ$ there is an additional internal diagram (fig.15.9)

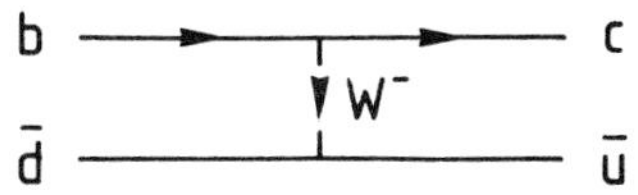


Fig.15.9

### 15.5 The Kobayashi-Maskawa matrix

In the first two generations the Cabibbo matrix is

$$\begin{pmatrix} d' \\ s' \end{pmatrix} = \begin{pmatrix} \cos\theta_c & \sin\theta_c \\ -\sin\theta_c & \cos\theta_c \end{pmatrix} \begin{pmatrix} d \\ s \end{pmatrix} \tag{15.5}$$

We might add a weak quark current $G(\bar{t}\gamma_\mu b)$ but since the $b$ quark decays across the generation gap a little bit of $b$ must be mixed into $d$, $s$. [Incidentally, writing the $-\frac{1}{3}$ charge quark states as mixtures as opposed to the charge $+\frac{2}{3}$ quark states is purely a matter of convention.] Cabibbo mixing can be represented by

$$\begin{pmatrix} d' \\ s' \end{pmatrix} = U_c \begin{pmatrix} d \\ s \end{pmatrix} \tag{15.6}$$

where $U_c$ is a real, one parameter unitary matrix—$U_c^+ U_c = 1$. The obvious generalisation is

$$\begin{pmatrix} d' \\ s' \\ b' \end{pmatrix} = U_{KM} \begin{pmatrix} d \\ s \\ b \end{pmatrix} \tag{15.7}$$

where $U_{KM}^+ U_{KM} = 1$ and $U_{KM}$ is a $3 \times 3$ unitary matrix—the Kobayashi-Maskawa matrix. This is very like a rotation matrix in three dimensions, so we might anticipate three real parameters. In fact the most general form of such a matrix has 4 parameters—three rotations and a phase, to be located within the matrix somewhere, which cannot be defined away by arbitrarily changing the phases of the quark states [see Georgi – *Weak Interactions and Modern Particle Theory*, page 51, or Commins and Bucksbaum – *Weak Interactions of Leptons and Quarks*, §4.3]. This may be important, for an imaginary piece in the matrix means $T$ (time reversal) violating weak interactions and hence, if the $CPT$ theorem is true, $CP$ violation in the weak interactions. In fact the long lived $K_L^\circ$ decays to the $CP = -1$ state $\pi^+\pi^\circ\pi^-$ but has a branching ratio $\sim 10^{-3}$

into the $CP = +1$ state $\pi^+\pi^-$. This has been known since 1964, described, and can now be accomodated within the K-M matrix (but not within the Cabibbo matrix) but is NOT understood. At present all we can say is that the quark couplings, as determined experimentally, are consistent with a (unitary) K-M matrix. We have no idea of the physics underlying the mixing.

### 15.6 The GIM mechanism

The introduction of the second weak doublet $\binom{c}{s'}$ was motivated by a remarkable fact of particle physics. At the tree level there seem to be no flavour changing neutral weak currents, and at the loop level change of flavour without change of charge is very small. This is best established for strange particles: for example, the branching ratio for $K^\circ \to \mu^+\mu^- \sim 10^{-8}$ and $K^\circ$ oscillates into $\bar{K}^\circ$ rather slowly. Thus the diagrams

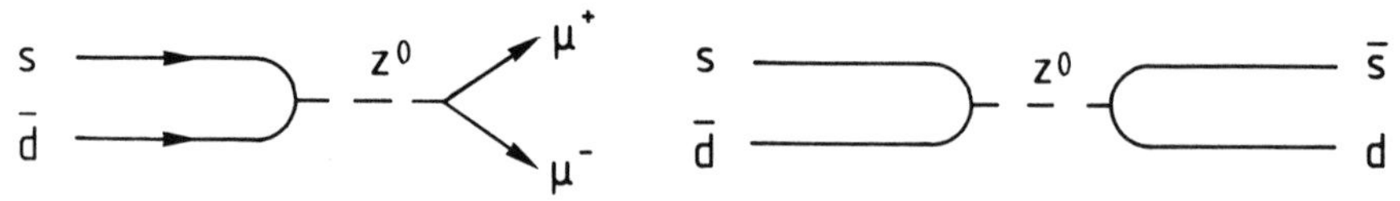


(a) (b)

Fig.15.10

have negligible couplings and the box diagrams such as

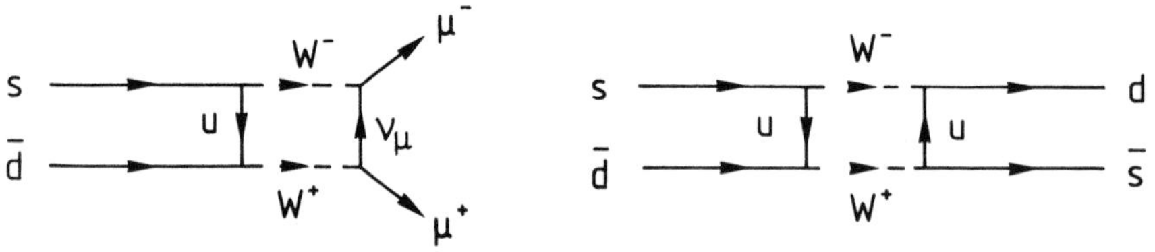


(a) (b)

Fig.15.11

give far too high rates.

Suppose the $Z^\circ$ couples to $(s's')$, $(d'd')$. Then

$$\begin{aligned}(s's') + (d'd') &= (s\cos\theta_c - d\sin\theta_c)^2 + (s\sin\theta_c + d\cos\theta_c)^2 \\ &= (ss) + (dd)\end{aligned} \tag{15.8}$$

and the right hand side of (15.8) contains no terms in $(sd)$. The $Z^\circ$ decouples from $(sd)$ and this removes the tree diagrams in fig.15.10. The loop diagrams of fig.15.11 are more complicated. With a $\binom{c}{s'}$ weak doublet, we replace the box section in fig.15.11 by the sum of two sections, fig.15.12.

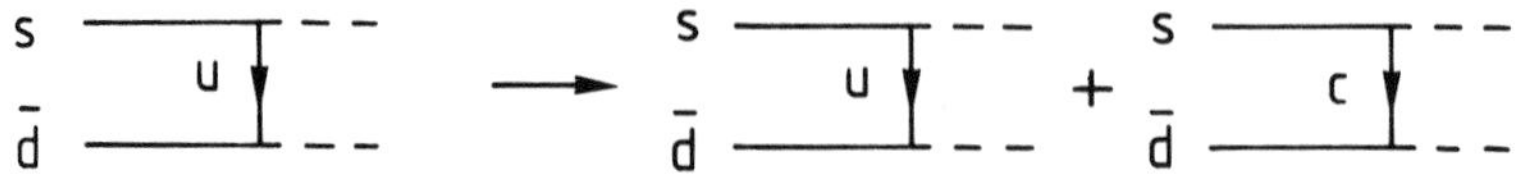


Fig.15.12

Thus

$$(su)(ud) \to (su)(ud) + (sc)(cd)$$

and the sum of the products of couplings yields

$$\sin\theta_c \cos\theta_c + \cos\theta_c(-\sin\theta_c) = 0 \tag{15.9}$$

With universal couplings of $W^{\pm}$ to the weak doublets $\binom{u}{d'}$, $\binom{c}{s'}$ the sum of the two amplitudes on the right hand side of fig.15.12 vanishes in the limit where $m_c = m_u$. The interactions leading to $K^\circ \to \mu^+\mu^-$, $K^\circ \to \bar{K}^\circ$ in fact only take place because the $c$ quark is heavier than the $u$ quark, and the $c$ quark propagator makes the second square ended diagram different in magnitude from the first, for moderate four-momentum transfer. Obviously the $c$ quark mass cannot be too great, or the famous GIM (Glashow-Illiopoulos-Maiani) mechanism we have described would not adequately suppress second order flavour changing neutral current processes. The existence of the $c$ quark was confidently predicted several years before its discovery, and its approximate mass was also predicted from the rate at which $K^\circ$ oscillates with $\bar{K}^\circ$. We have not considered how to calculate loop diagrams such as those shown in fig.15.11, but the form of the result can be obtained on little more than dimensional grounds.

Consider the rate at which $K^\circ$ oscillates into $\bar{K}^\circ$. (The mass difference of the mass eigenstates $K_L^\circ$ and $K_S^\circ$ is directly related to this rate.) The matrix element $< K^\circ|W|\bar{K}^\circ >$ must have the dimensions of mass. Fig.11.15(b) yields ingredients

$$g^4 \sin^2\theta_c \cos^2\theta_c \frac{1}{(q^2+M_W^2)^2}\frac{1}{(q^2+m_u^2)^2}|\psi(0)|^2 \tag{15.10}$$

where conserved four-momentum $q_\mu$ runs round the loop and for large $q^2$ physical momentum transfer is neglected. $\psi(0)$ is an effective quark wave function in the region of the origin. If we neglect $m_u$ and extract a factor of $M_W^4$, (15.10) becomes

$$\sim G_F^2 \sin^2\theta_c \cos^2\theta_c \frac{1}{\left(1+\frac{q^2}{M_W^2}\right)^2}\frac{1}{q^4}|\psi(0)|^2 \tag{15.11}$$

This must be integrated, with appropriate weight, over all $q^2$. Since we shall have

$$|\psi(0)|^2 \approx M_K^3 \tag{15.12}$$

and the dimensions of $G_F^2$ are $\mathrm{GeV}^{-4}$, then the factor

$$\int \frac{1}{\left(1+\frac{q^2}{M_W^2}\right)^2}\frac{1}{q^4}dW(q^2) \tag{15.13}$$

must have dimensions $\mathrm{GeV}^2$ in order that $< K^\circ|W|\bar{K}^\circ >$ has dimensions GeV. Then (15.13) must become

$$\approx \int \frac{dq^2}{\left(1+\frac{q^2}{M_W^2}\right)^2} \approx M_W^2 \tag{15.14}$$

and

$$< K^\circ|W|\bar{K}^\circ > \approx G_F^2 \sin^2\theta_c \cos^2\theta_c M_K^3 M_W^2 \tag{15.15}$$

The essential point is that the cube of a small mass, $\sim M_K$, is expected from the quark wave functions, and the only scale available for insertion of the other two powers of mass is $M_W$, if we neglect the $c$ quark.

With the $c$ quark included, the factor of

$$\frac{1}{(q^2+m_u^2)^2}$$

in (15.9) is replaced

$$\begin{aligned}\frac{1}{(q^2+m_u^2)^2} &\to \left(\frac{1}{q^2+m_u^2}-\frac{1}{q^2+m_c^2}\right)^2 \\ &\approx \frac{1}{q^4}\frac{1}{\left(1+\frac{q^2}{m_c^2}\right)^2}\end{aligned} \tag{15.16}$$

Comparing (15.16) with (15.10) and (15.11) it is evident that provided $M_W^2 \gg m_c^2 \gg m_u^2$ then the integral over weighted intermediate states is now cut off by $m_c^2$ rather than $M_W^2$, and that (15.15) is replaced by

$$< K^\circ|W|\bar{K}^\circ > \approx G_F^2 \sin^2\theta_c \cos^2\theta_c M_K^3 m_c^2 \tag{15.17}$$

Equation (15.17) is not sufficiently accurate to allow us to make our own estimate of an upper limit on $m_c$, for we have been unable to keep track of factors of $2\pi$ and the factor $M_K^3$ could equally well have been chosen as $m_\pi^3$ or $(0.5\text{ fm})^{-3}$. We have however identified the essential ingredients in a dimensionally correct expression and seen how introduction of the $c$ quark lines replaces a factor $M_W^2$ by $m_c^2$. Equation (15.17) yields

$$< K^\circ|W|\bar{K}^\circ > = 5\times 10^{-13}\text{ GeV} = 5\times 10^{-4}\text{ eV}$$

if $m_c \sim 1$ GeV, whereas (15.15) yields 5 eV. The difference in mass between the two mass eigenstates is twice this matrix element. Experimentally

$$M_{K_L^\circ} - M_{K_S^\circ} = 3.5\times 10^{-6}\text{ eV}$$

The evaluation of the integral over intermediate states yields the proper factors of $2\pi$ and the quantity we have regarded as $|\psi(0)|^2$ can be estimated from the decay rate of the charged kaon into $\mu\nu_\mu$. The predicted mass difference [Gaillard and Lee, *Phys. Rev.* **D10** 897 (1974)] is

$$\begin{aligned} m_{K_L^\circ} - m_{K_S^\circ} &\approx \frac{G_F^2}{4\pi^2}\sin^2\theta_c\cos^2\theta_c f_K^2 M_K m_c^2 \\ &\approx 2\times 10^{-6} m_c^2\text{ eV}\end{aligned} \tag{15.18}$$

where $m_c$ is in units of GeV and $f_K \simeq m_\pi$. The prediction that the $c$ quark mass must be $\sim 1.5$ GeV was made from a form equivalent to (15.18), before the discovery of the $J/\psi$ in November, 1974.

The above discussion of the GIM mechanism and the prediction of the $c$ quark mass was made in the context of two generations. The scheme generalises to three generations with a $K - M$ mixing matrix, but while the existence of the strange quark implied the existence of charm, the existence of two generations of quarks did not imply the existence of a third.

The pattern of weak interactions among the quarks has a very simple description, but description is not understanding. The mixtures $d'$, $s'$, $b'$ which couple to the weak interactions have the same charge and colour as the mass eigenstates $d$, $s$, $b$. The eigenstates

$$\begin{matrix} u & c & t \\ d & s & b \end{matrix}$$

have huge mass splittings both within and between generations, and the splitting increases with generation number. The degree of mixing in the weak current states

$$\begin{matrix} u & c & t \\ d' & s' & b' \end{matrix}$$

falls rapidly as the generation number and gap increase. The weak intermediate bosons are far more massive than the five identified quarks, but the top quark could well have a mass $\sim M_W$. Perhaps the higher generations will one day be interpreted as some form of metastable states, but so far no one has made any sense of this curious pattern.

Since the weak interactions induce mixing among the quarks, it is natural to wonder whether mixing could also be induced among the three generations of lepton. This is possible, if all neutrino masses are not zero.

There are already stringent limits on the existence of neutrino oscillations, but there remains Davis' observation that the flux of neutrinos from the sun inducing $\nu_e + {}^{37}\mathrm{Cl} \to {}^{37}\mathrm{A}$ is only one third what it should be. Perhaps there is nothing wrong with the sun (or astrophysics) but that a $\nu_e$ departing the core of the sun is on average an equal mixture of $\nu_e$, $\nu_\mu$, $\nu_\tau$ when it arrives in the Homestake Mine, Lead, South Dakota.... Experiments setting limits on neutrino oscillations or determining better limits on neutrino masses are important.

### 15.7 Real intermediate bosons

It was of course the weak interactions of quarks which made visible the (real) $W^\pm$, $Z^\circ$ at the CERN $p\bar{p}$ collider, through a weak Drell-Yan mechanism; such as

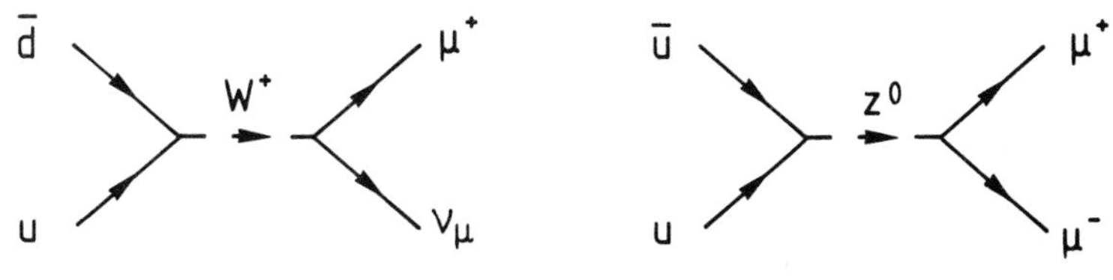


Fig.15.13

The top quark could in principle be seen through the related process

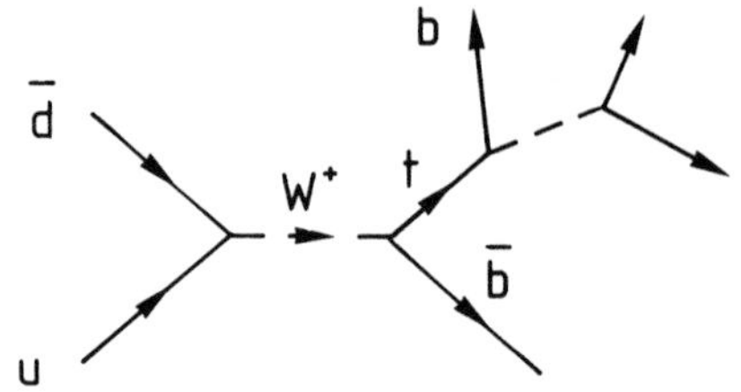


Fig.15.14

In 1983 the UA1 collaboration at the CERN collider had a cluster of 7 events consistent with this process for $m_t \sim 40$ GeV. More data and further studies of background have now resulted in a UA1 limit $m_t > 44$ GeV. The Fermilab collider is now yielding data at a $p\bar{p}$ centre of mass energy of 1800 GeV. Perhaps the top quark will appear at last.

## Problems

15.1 In fig.15.11(a) a diagram for $K° \to \mu^+\mu^-$ is shown, second order in the charged weak interaction. In fact $K° \to \mu^+\mu^-$ is possible through a first order charged interaction and electromagnetic effects. Find diagrams which are first order in the weak interactions and lead to $K° \to \gamma\gamma$. Examine the dimensions and physical content of the matrix element and make an estimate on dimensional grounds of the rate for $K° \to \gamma\gamma$. [Do not attempt to perform any loop integration.] Estimate the branching ratio for $K_L^\circ \to \gamma\gamma$ [$\tau_{K_L^\circ} = 5 \times 10^{-8}$s] and hence make an estimate of the branching ratio for $K_L^\circ \to \mu^+\mu^-$ via a first order (charged) weak interaction and electromagnetic terms. [This process is believed to dominate over second order weak interaction amplitudes.]

15.2 Find the second order charged weak interaction diagrams which mix $D°(c\bar{u})$ and $\bar{D}°(\bar{c}u)$. Estimate the mass difference between the two (mixed) eigenstates and hence the timescale for conversion of $D°$ into $\bar{D}°$. Is there any reasonable prospect of being able to study $D° - \bar{D}°$ mixing experimentally? [$\tau_{D°} = 4 \times 10^{-13}$s.]

15.3 Set up all the second order weak interaction amplitudes which induce $K° \to \bar{K}°$ and $K° \to K°$. Find the composition and masses of the (mixed) eigenstates and their mass difference in terms of the weak interaction matrix element. You should find states

$$K_1^\circ = \frac{1}{\sqrt{2}}(K° + \bar{K}°)$$

and

$$K_2^\circ = \frac{1}{\sqrt{2}}(K° - \bar{K}°)$$

[It is natural and conventional to choose arbitrary phases such that $K_1^\circ$ is $C = +1$ and $K_2^\circ$ $C = -1$.] These states are also CP eigenstates, and the shortlived state $K_S^\circ$ decays predominantly into two pions and must be predominantly $K_1^\circ$. Consider the likely decay modes of the longlived state $K_L^\circ$ which is predominantly $K_2^\circ$ in composition. [If all interactions conserve both CP and CPT, the $K_S^\circ$ and $K_L^\circ$ would be identical with $K_1^\circ$ and $K_2^\circ$, and their decay modes would be circumscribed. The state $K_L^\circ$ decays with a branching ratio of $2 \times 10^{-3}$ into $\pi^+\pi^-$ and $9 \times 10^{-4}$ into $\pi^\circ\pi^\circ$. It is either a state of mixed CP or CP is violated in the decay process, or both. The phenomenon of CP violation is well established in the $K^\circ - \bar{K}^\circ$ system, but has so far appeared nowhere else. [$\tau_{K_S^\circ} = 0.89 \times 10^{-10}$s, $\tau_{K_L^\circ} = 5.18 \times 10^{-8}$s.]

15.4 Ignoring CP violation, find the distribution with respect to proper time for the appearance of $\pi^+\pi^-$ in a beam consisting initially of $K^\circ$. Then find the distribution of final states $\pi^+e^-\bar{\nu}_e$ and $\pi^-e^+\nu_e$ in beams consisting initially of a) $K^\circ$ and b) $\bar{K}^\circ$.

An almost pure beam of $K_L^\circ$ can be prepared simply by going far enough downstream of the target. When such a beam is passed through matter $K_S^\circ$ are regenerated. Explain as quantitatively as possible.

15.5 Consider the processes which may convert $B^\circ(b\bar{d})$ into $\bar{B}^\circ(\bar{b}d)$ and $B_s^\circ(b\bar{s})$ into $\bar{B}_s^\circ(\bar{b}s)$. Try and make some estimate of the rate at which $B^\circ$ will oscillate into $\bar{B}^\circ$ and $B_s^\circ$ into $\bar{B}_s^\circ$. Why are you not asked to consider processes such as $B^\circ \to B_s^\circ$?

Consider what the experimental signature of such oscillations might be. It has recently been claimed that there is evidence for oscillations in the $B^\circ - \bar{B}^\circ$ system, as opposed to the $B_s^\circ - \bar{B}_s^\circ$ system where the effect should be easier to observe. Why is it expected that oscillations will be faster in the $B_s^\circ - \bar{B}_s^\circ$ system. Can you draw any conclusions about the top quark mass from observation of oscillations in the neutral $B$ systems? [$\tau_b \sim 1.2 \times 10^{-12}$s] Reference: P.J. Franzini *Phys. Rep.* **173** 1 (1989)

[If bottom oscillations can be observed with precision, it is possible tht these systems might provide a new arena for the exhibition of CP violation. There is considerable interest at present in the construction of dedicated $B$ factories. You might care to consider the desirable features of such factories.]

## 16. THE (UN)UNIFIED ELECTROWEAK INTERACTION.

### 16.1 Weak isospin

There are two marked similarities between the weak and electromagnetic interactions. Both are mediated by spin 1 fields and the couplings are very similar. However, the charged weak interaction violates parity maximally, coupling only to the left handed component of fermions, and the intermediate bosons $W^{\pm}$, $Z^{\circ}$ have masses $\sim 100$ GeV whereas the photon mass is $\sim 0$ ($< 10^{-15}$eV). The idea of a unified electroweak theory is nonetheless attractive. The standard model (Glashow-Weinberg-Salam) in fact entangles the two ($W^{\pm}$ are charged) but a true unification has not been accomplished.

The $W^{\pm}$ act between pairs of left handed particles

$$\begin{pmatrix} \nu_{e_L} \\ e_L \end{pmatrix}, \quad \begin{pmatrix} u_L \\ d'_L \end{pmatrix} \quad \cdots$$

It is an attractive idea to suppose that these pairs are doublets of a weak isospin, and that the interactions between them are mediated by a weak isospin triplet ($W^{\pm}$, $W^{\circ}$), just as the pion couples nucleons, and possibly a weak isospin singlet ($\gamma$?).

There are (obviously) several things badly wrong with this idea. For a perfect $SU(2)$ the masses of the two members of a given doublet should be identical, the members of the triplet should have the same mass, and an isospin singlet should couple with equal strength to the two members of a doublet, yet the neutrino has no electric charge and $m_e/m_\nu > 10^4$. There is also the problem of the right handed components of the fermions. If the neutrinos are massless, right handed neutrinos could be excluded and the state $e_R$ would be a weak isospin singlet, with no coupling to the $W$s. In the quark sector, $u_R$, $d'_R$ would both have to be singlets and it begins to get ugly.

If there is an underlying weak isospin, it has to be broken, and the photon cannot be a singlet because it couples to $e_L$ and not to $\nu_e$. This can be achieved if the physical photon is an appropriate mixture of an isospin triplet and singlet. Such a mixture can only result from breaking weak isospin $SU(2)$ symmetry. Suppose weak $SU(2)$ is broken in such a way as to mix a weak isospin triplet member $W^{\circ}$ and a singlet $\Gamma$. The two new orthogonal states are

$$\begin{aligned} Z^{\circ} &= W^{\circ} \cos\theta_W - \Gamma \sin\theta_W \\ \gamma &= W^{\circ} \sin\theta_W + \Gamma \cos\theta_W \end{aligned} \tag{16.1}$$

(where $\theta_W$ is the Weinberg angle).

In this weak isospin picture members of a doublet $\binom{i}{j}$ interact via exchange of $W^{\pm,0}$ in the familiar way (fig.16.1)

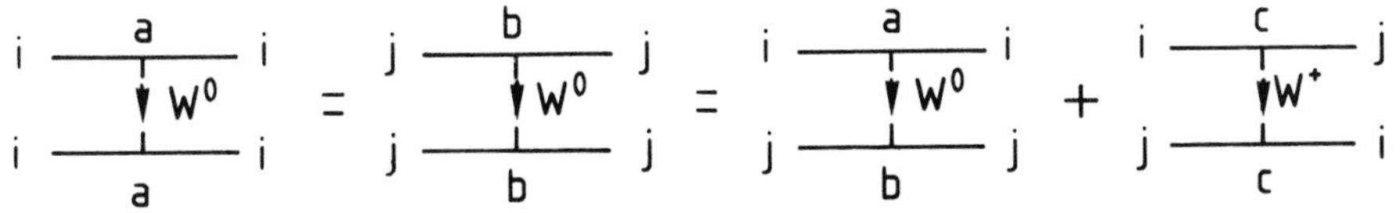


Fig.16.1

As usual, $a^2 = b^2 = ab + c^2 \ldots$ set $a = -b$, $c = \sqrt{2}b$ and $b = g$. The $W^\circ$ couplings are thus related to the charged coupling $\sqrt{2}g$. An isosinglet exchange ($a = b$, $c = 0$) has just one coupling $g'$ to both members of a doublet, but the value of $g'$ may differ from doublet to doublet. Because of the experimental universality of the charged weak interaction, take the constant $g$ to be universal. Let the isosinglet couplings be multiples of a single coupling; $g'Y$ where $Y$ is multiplet dependent.

The first generation fermion multiplets are

$$\begin{array}{cccccc} T_3 & & T = \dfrac{1}{2} & & T_3 & T = 0 \\ & & & & & \\ \begin{array}{c} +1/2 \\ -1/2 \end{array} & \begin{pmatrix} \nu_{eL} \\ e_L^- \end{pmatrix} & , & \begin{pmatrix} u_L \\ d'_L \end{pmatrix} & ; \quad 0 & e_R^-, u_R, d'_R \end{array} \tag{16.2}$$

The electron has charge $-e$ and the neutrino charge zero, so

$$\begin{aligned} 0 &= a \sin\theta_W + g'Y_\ell \cos\theta_W \\ -e &= b \sin\theta_W + g'Y_\ell \cos\theta_W \end{aligned} \tag{16.3}$$

With $b = g$ and the isotriplet coupling relation $a = -b$

$$\begin{aligned} 0 &= -g \sin\theta_W + g'Y_\ell \cos\theta_W \\ -e &= g \sin\theta_W + g'Y_\ell \cos\theta_W \end{aligned} \tag{16.4}$$

where $Y_\ell$ is characteristic of the left handed lepton doublet. Choose $Y_\ell = -1$ so that as $\theta_W \to 0$, $g' \to e$. Then

$$0 = -g \sin\theta_W - g' \cos\theta_W \qquad \frac{g'}{g} = -\tan\theta_W \tag{16.5}$$

$$-e = 2g \sin\theta_W \tag{16.6}$$

Extend the argument to the (left handed) quark doublets, with the singlet $\Gamma$ coupling $g'Y_q$:

$$\begin{aligned} +\frac{2}{3}e &= -g \sin\theta_W + g'Y_q \cos\theta_W \\ -\frac{1}{3}e &= g \sin\theta_W + g'Y_q \cos\theta_W \end{aligned} \tag{16.7}$$

Then subtracting we obtain

$$e = -2g \sin\theta_W$$

which is also the relation obtained for the lepton doublet, and reflects the unit separation of electric charge of the members.
Adding the two eqs.(16.7)

$$\frac{1}{3}e = 2g' \cos\theta_W Y_q = eY_q \quad \therefore \quad Y_q = +\frac{1}{3}$$

Noting that the electric charges and the eigenvalues of $T_3$ are separated in a doublet by one unit in both cases, we can write a general expression

$$\begin{aligned} Q &= -2T_3 g \sin\theta_W + g'Y\cos\theta_W \\ &= e(T_3 + \frac{Y}{2}) \end{aligned} \tag{16.8}$$

$$Y = -1 \quad \begin{pmatrix} \nu \\ e \end{pmatrix}_L$$

$$Y = +\frac{1}{3} \quad \begin{pmatrix} u \\ d' \end{pmatrix}_L$$

For $e_R$, $u_R$, $d'_R$, $T_3 = 0$ and $Y$ must be tuned singlet by singlet to yield the correct charge:

$$Y(e_R) = -2,\ Y(u_R) = +\frac{4}{3},\ Y(d'_R) = -\frac{2}{3}.$$

This looks horrid and *ad hoc*, but did $\nu_R$ exist it would have $T_3 = 0$ (to decouple $W$s) and $Y = 0$ to decouple the photon. Then $Y(\nu_R) - Y(e_R) = 2$; $Y(u_R) - Y(d'_R) = 2\ldots$

Despite the tuning by hand of the hypercharges $Y$ in order to match the electromagnetic couplings of quarks and leptons, the scheme has merit because with the hypercharges identified the $Z^\circ$ couplings are predicted and these can be determined experimentally.

With

$$Z^\circ = W^\circ \cos\theta_W - \Gamma\sin\theta_W$$

and a $W^\circ$ coupling $-2gT_3$, $\Gamma$ coupling $g'Y$, the $Z^\circ$ coupling to fermions is written

$$-2T_3 g\cos\theta_W - g'Y\sin\theta_W \tag{16.9}$$

With the relation $g'/g = -\tan\theta_W$ (16.5) the $Z^\circ$ couplings become

$$g\cos\theta_W(-2T_3 - Y\frac{g'}{g}\tan\theta_W) = g\cos\theta_W(-2T_3 + Y\tan^2\theta_W) \tag{16.10}$$

Rewrite (16.10) in terms of $T_3$ and the charge $Q$, using (16.8)

$$\frac{Q}{e} = T_3 + \frac{Y}{2}$$

Then

$$g\cos\theta_W\left(-2T_3 + \left(2\frac{Q}{e} - 2T_3\right)\tan^2\theta_W\right) = \frac{2g}{\cos\theta_W}\left(-T_3 + \frac{Q}{e}\sin^2\theta_W\right) \tag{16.11}$$

The $T_3$ operator must project out left handed doublets and give zero applied to the right handed singlets. A factor of $(1 - \gamma_5)$ kills the RH pieces explicitly, so the full $Z^\circ$ interaction thus contains the operator

$$\frac{g}{\cos\theta_W}\left(T_3\gamma_\mu(1-\gamma_5) - 2\frac{Q}{e}\sin^2\theta_W\gamma_\mu\right) \tag{16.12}$$

[$\frac{1}{2}(1-\gamma_5)$ is the left hand spin projection operator.]
Eq.(16.12) is conventionally rewritten in the form

$$\frac{g}{\cos\theta_W}(c_V\gamma_\mu - c_A\gamma_\mu\gamma_5) \tag{16.13}$$

and is of course to be sandwiched between the appropriate pair of spinors. Instead of the $V - A$ coupling of the $W^\pm$, the $Z^\circ$ has couplings $c_V V - c_A A$.

The axial coupling is contributed only by the $W^\circ$ component of the $Z^\circ$ and is universal: the vector coupling is a mixture of $W^\circ$ and $\Gamma$. The values of $c_V$ and $c_A$ can be written down at once.

**Table 16.1**

| | $T_3$ | $Q/e$ | $c_V$ | $c_A$ | |
|---|---|---|---|---|---|
| $\nu_e$ | $+\frac{1}{2}$ | $0$ | $+\frac{1}{2}$ | $+\frac{1}{2}$ | $V - A$ of course |
| $e^-$ | $-\frac{1}{2}$ | $-1$ | $-\frac{1}{2} + 2\sin^2\theta_W$ | $-\frac{1}{2}$ | |
| $u$ | $+\frac{1}{2}$ | $+\frac{2}{3}$ | $+\frac{1}{2} - \frac{4}{3}\sin^2\theta_W$ | $+\frac{1}{2}$ | |
| $d'$ | $-\frac{1}{2}$ | $-\frac{1}{3}$ | $-\frac{1}{2} + \frac{2}{3}\sin^2\theta_W$ | $-\frac{1}{2}$ | |

The same pattern is repeated for the second and third generations. Note that as $\sin^2\theta_W \to 0 \quad Z^\circ \to W^\circ$ and the neutral weak couplings take on the $V - A$ form of the $W^\pm$ couplings.

The electromagnetic and weak interactions have been entangled, but unification has not been achieved because the theory contains the two parameters $g$ and $g'$, or equivalently $e$ and $\sin^2\theta_W$. The value of $\sin^2\theta_W$ is NOT predicted and must be measured: $\sin^2\theta_W \simeq 0.22$. (For the electron, $c_V \simeq 0$ and the weak neutral current is nearly pure axial vector.) A unified theory would predict the value of $\sin^2\theta_W$.

The astonishing thing about this hideous edifice is that it works. The electroweak interactions of leptons and quarks are admirably described by the couplings we have derived, and the value of $\sin^2\theta_W$ is universal.

The $Z^\circ$ couplings, and hence $\sin^2\theta_W$, have been obtained experimentally in a number of ways:

(1) $\left.\begin{matrix}\nu_\mu e \to \nu_\mu e \\ \bar{\nu}_\mu e \to \bar{\nu}_\mu e\end{matrix}\right\}$ mediated ONLY by $Z^\circ$ exchange

(2) $\bar{\nu}_e e \to \bar{\nu}_e e$ (ANTI neutrinos from fission reactors)

(3) $\nu_\mu N \to \nu_\mu X$

(4) Parity violation in deep inelastic scattering of polarised electrons from deuterium (a famous SLAC experiment). The (tiny) parity violation is an interference between the diagrams

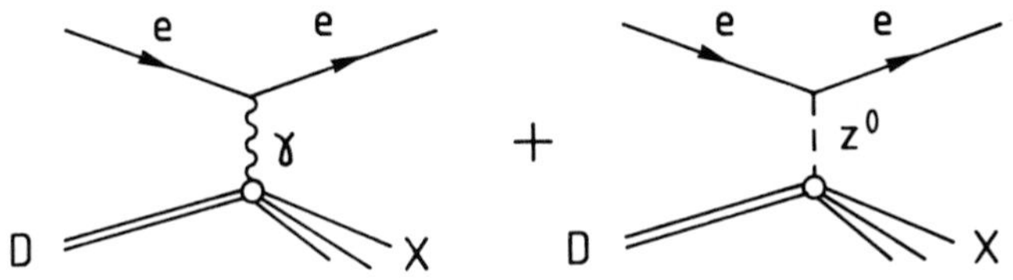


Fig.16.2

(5) Forward-backward asymmetry in

$$e^+e^- \to \mu^+\mu^-(\tau^+\tau^-)$$
$$e^+e^- \to b\bar{b} \text{ (JADE at PETRA)}$$
$$e^+e^- \to c\bar{c}$$

In (5) above the asymmetry is again due to $\gamma - Z^\circ$ interference (fig.16.3)

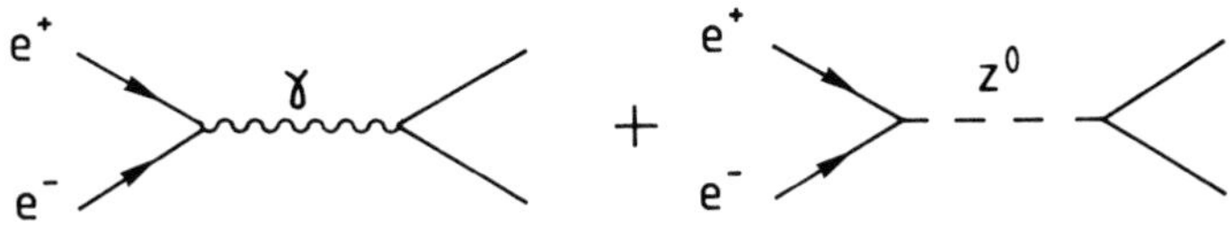


Fig.16.3

The familiar angular distribution characteristic of a vector interaction between spin $\frac{1}{2}$ particles

$$\frac{d\sigma}{d\Omega} \propto 1 + \cos^2\theta$$

is modified by interference between the two amplitudes of fig.16.3 to yield

$$\frac{d\sigma}{d\Omega} \propto 1 + a\cos\theta + \cos^2\theta$$

at centre of mass energies $\sim$ 35 GeV. The coefficient $a$ depends on the $Z^\circ$ couplings at the two vertices and on the mass of the $Z^\circ$. The asymmetric term comes about because the $L$ and $R$ couplings are unbalanced in the sum of the two amplitudes, and is thus a manifestion of parity violation in the neutral weak interaction. [A forward-backward asymmetry of this kind is not necessarily a signature of parity violation: $e^+e^- \to e^+e^-$ is enormously forward peaked because of scattering through exchange of soft photons carrying momentum.] It is not difficult to work out the coefficients $a$, given the $Z^\circ$ couplings (see problem 16.5).

### 16.2 Mixing and mass

The mixing between $W^\circ$ and $\Gamma$ explicitly violates weak isospin. One might envisage some Hamiltonian operator coupling the $W$s and $\Gamma$ to, for example, a doublet of scalar bosons and one might represent the effect schematically as

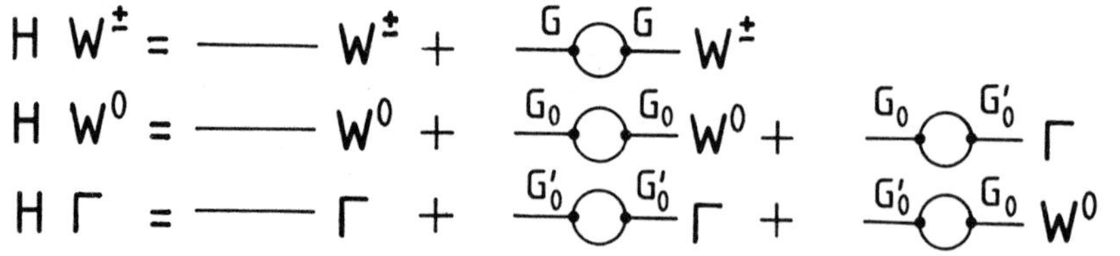

$$H\,W^{\pm} = \,\underline{\qquad}\, W^{\pm} + \overset{G}{\underset{G}{\bigcirc}} W^{\pm}$$

$$H\,W^{0} = \,\underline{\qquad}\, W^{0} + \overset{G_0}{\underset{G_0}{\bigcirc}} W^{0} + \overset{G_0}{\underset{G_0'}{\bigcirc}} \Gamma \qquad (16.14)$$

$$H\,\Gamma = \,\underline{\qquad}\, \Gamma + \overset{G_0'}{\underset{G_0'}{\bigcirc}} \Gamma + \overset{G_0'}{\underset{G_0}{\bigcirc}} W^{0}$$

where the doublet somehow violates weak $SU(2)$.

If we started with the $W$ and $\Gamma$ boson masses zero and the loops contribute the masses as well as mixing $W^{\circ}$ and $\Gamma$

$$M_{W^{\pm}} = G^2$$

(where $G^2$ contains the couplings and other factors). The coupled equations derived from (16.14)

$$MW^{\circ} = G_0^2 W^{\circ} + G_0 G_0' \Gamma$$
$$M\Gamma = G_0'^2 \Gamma + G_0 G_0' W^{\circ} \qquad (16.15)$$

must be diagonalised. The eigenvalues are easy:

$$\begin{vmatrix} M - G_0^2 & -G_0 G_0' \\ -G_0 G_0' & M - G_0'^2 \end{vmatrix} = 0$$

$$(M - G_0^2)(M - G_0'^2) - G_0^2 G_0'^2 = 0$$
$$M(M - G_0^2 - G_0'^2) = 0$$

Then

$$M_{\gamma} = 0 \quad , \quad M_{Z^{\circ}} = G_0^2 + G_0'^2$$

If $G_0^2 = G^2$ which is nice because in the absence of mixing $W^+$, $W^-$, $W^{\circ}$ would have the same mass, then $M_{Z^{\circ}} > M_{W^{\pm}}$

The evolution equations for elementary bosons are in fact second order rather than first order, and we might expect this picture to work better with the masses replaced by their squares. In that case,

$$M_{Z^{\circ}}^2 = M_{W^{\pm}}^2 (G^2 + G_0'^2)/G^2 \qquad (16.16)$$

Regardless of whether a mass or mass squared operator is appropriate, the mixing of the two states is easily obtained. The state with zero mass is

$$\gamma = W_0 \sin\theta_W + \Gamma \cos\theta_W$$

Then

$$\hat{M}^{(2)}\gamma = \hat{M}^{(2)}(W_0 \sin\theta_W + \Gamma \cos\theta_W) = 0 \qquad (16.17)$$

Therefore

$$G_0^2 \sin\theta_W + G_0 G_0' \cos\theta_W = 0 \qquad \text{(coefficient of } W^{\circ})$$
$$G_0 G_0' \sin\theta_W + G_0'^2 \cos\theta_W = 0 \qquad \text{(coefficient of } \Gamma) \qquad (16.18)$$

$$\tan\theta_W = -\frac{G_0'}{G_0}$$

Then

$$\frac{M_{Z^\circ}^2}{M_W^2} = \frac{G^2 + G_0'^2}{G^2} = \frac{1}{\cos^2\theta_W} \quad ; \quad M_{Z^\circ} = \frac{M_W}{\cos\theta_W} \tag{16.19}$$

The electron charge $e$ and the (measured) value of $\sin^2\theta_W$ determine $g$ and hence the charged current $W^\pm$ coupling, $\sqrt{2}g$. The value of $\sqrt{2}g$, together with the Fermi coupling constant $G_F$, determines the mass of $W^\pm$: $M_W \simeq 80$ GeV. With $\sin^2\theta_W = 0.22$, $M_{Z^\circ} \simeq 90$ GeV. These values were predicted before the observation of $W^\pm$, $Z^\circ$ at the CERN SPS $p\bar{p}$ collider. The scheme we have sketched may seem crazy, but this is what nature has chosen to do.

There is a theoretical imperative for such a scheme. A theory embodying elementary massive fermions and massive spin 1 intermediate bosons remains plagued by divergences. A massless theory, elegant because of the muddied chiral properties of the weak interactions, is not. A renormalisable electroweak theory embodying gauge invariance can be constructed by including scalar bosons present in the vacuum. Weak isospin is broken, in the simplest such model, by having only the $T_3 = -\frac{1}{2}$ member of a weak isospin doublet with non-zero vacuum expectation value. Interaction with this Higgs field is used to generate masses and mixing, but the masses are still inserted by hand. The emergent scheme is as we have described.

### 16.3 The Weinberg-Salam model — unexplained

Suppose we want a theory which embodies weak isospin and the hypercharge interactions, and which starts with massless vector bosons. If the underlying theory is invariant under the operations of weak isospin, $SU(2)_L$, then the fermions must also have zero mass, for it is the mass term in the Dirac equation which couples left to right handed fermions. The particles in the theory can be made to grow masses by coupling them to a field already present in the vacuum. If this field is a Lorentz scalar, then the massless Dirac equation would be modified by such a background scalar potential

$$\gamma_\mu \partial_\mu \psi = 0 \rightarrow \{\gamma_\mu \partial_\mu + ig < \phi >\}\psi = 0 \tag{16.20}$$

and the effective fermion mass would be $g < \phi >$, where $< \phi >$ is the vacuum expectation value of the new scalar field and $g$ is its coupling to the fermion field $\psi$. Since the mass term links $\psi_R$ (a weak singlet) to $\psi_L$ (a weak doublet with $T_3 = -\frac{1}{2}$) $< \phi >$ must, for this application, be the lower part of a weak doublet with $Y = 1$. The field $< \phi >$ is then electrically neutral, and the fermion singlet and doublet differ in $Y$ by one unit. The coupling $g$ has to be tuned, fermion by fermion, to produce the correct mass.

The $W$ and $Z$ masses and mixing can also be generated by coupling them to this vacuum field. In electromagnetism, the classical four-potential satisfies an equation

$$\Box A_\mu = 4\pi j_\mu \tag{16.21}$$

and if the four-current $j_\mu$ is proportional to $A_\mu$ the field is screened and becomes of finite range.

The classic example is the exponential attenuation of a magnetic field in a superconductor, where a potential macroscopic current exists in the form of Cooper pairs. The photon acquires an effective mass through the process

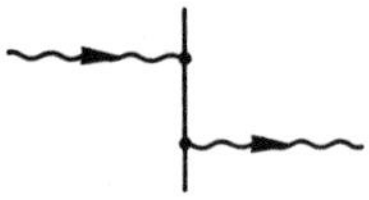

Fig.16.4

We therefore envisage an induced weak current in the vacuum, $<\phi>$ responding to the vector fields, and (16.14) becomes

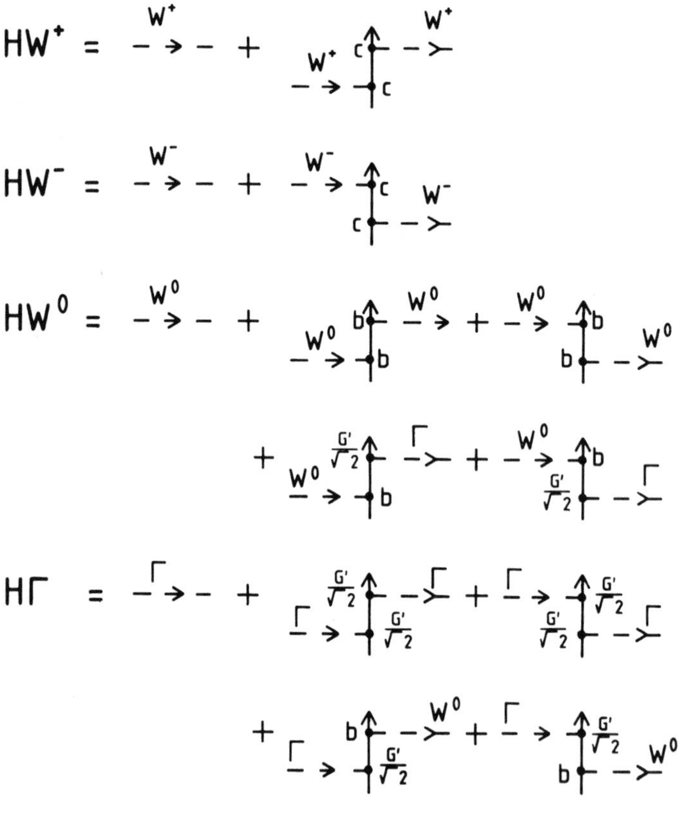


(16.22)

where the vertical line denotes a robust screening current. Interpreting the relations implied by (16.22) in terms of mass squared rather than mass, as suggested by (16.21), the relations (16.16–16.19) follow at once on noting that $c^2 = 2b^2 = G^2$, and the squared masses are proportional to the vacuum expectation value of the field $<\phi>$.

It is possible to arrange that the underlying theory is $\mathrm{SU}(2)_L$ invariant, yet the vacuum is not. The doublet $\phi$ must be self interacting in such a way that the lowest energy state of the field $\phi$ — the vacuum — does not correspond to zero field. The effect is known as spontaneous symmetry breaking and this mechanism for giving mass to initially massless states by tinkering with the vacuum is called the Higgs mechanism. Excitation of $< \phi >$ corresponds to production of physical scalars, Higgs scalars, which are needed to control otherwise divergent $WW$ processes. There is at present no evidence whatsoever for their existence.

The proper theory is constructed within the framework of local gauge invariance, an enormously powerful symmetry principle which dictates the structure we have only sketched. The peculiar virtues of such a framework are that first a great deal of floundering around searching for a consistent theory is eliminated, secondly such theories are renormalisable order by order in perturbation theory and therefore have, at least in principle, predictive power. There is however no unique choice for the $SU(2)_L$ properties of the background field and the self-interaction is put in by hand. It works, but there is a strong smell of condensed matter physics about the whole business and the true nature of the physics underlying this remarkably successful model is obscure.

The curious idea of the vacuum, the ground state, containing an expectation value for some field $\phi$ which, under present conditions, is non-zero has spawned developments in theoretical cosmology, the inflationary scenarios. In these models cooling of the infinitesimally early universe is accompanied by a phase change from a vacuum in which $< \phi >$ is zero to a lower energy state in which a vacuum expectation value develops. The latent heat drives an enormously rapid expansion — inflation — and such a scheme could account for the remarkably homogeneous universe we perceive. It is claimed that such models also yield a value for the parameter $\Omega$ indistinguishable from unity, $\Omega$ being the ratio of the mass density of the universe to that required to just close the universe. Luminous matter apparently contributes a value $\sim 0.01$, but the dynamics of galaxies and clusters of galaxies suggests $\Omega \gtrsim 0.1$.

It is clear that in the entangled electro-weak theory we have a new phenomenology. Perhaps the $W$s really are elementary and the vacuum asymmetric ... or perhaps the $W$s are composites of some unknown kind, in which case divergences in a theory of elementary massive vector bosons would not be a matter for concern. The new phenomenology of leptons, quarks, and bosons $W^{\pm}$, $Z^{\circ}$, $\gamma$ needs deeper (experimental) investigation. We must probe distances $\ll 10^{-16}$ cm, corresponding to (useful) energies $> 100$ GeV. It is here that the solutions to the real mysteries must be sought.

### Problems

16.1 Below eq.(16.19) we asserted that from the electron charge $e$, the value of $\sin^2 \theta_W$ and the value of $G_F$ the values of $M_{W^{\pm}}$ and $M_{Z^{\circ}}$ can be calculated. Verify that the results

$$M_{W^{\pm}} \sim 80 \text{ GeV} \qquad M_{Z^{\circ}} \sim 90 \text{ GeV}$$

are indeed obtained, with the minimal $T = \frac{1}{2}$ Higgs field.

16.2 Calculate the cross sections for

$$\nu_\mu e^- \rightarrow \nu_\mu e^-$$
$$\bar{\nu}_\mu e^- \rightarrow \bar{\nu}_\mu e^-$$

for centre of mass energies restricted by

$$m_e^2 \ll s \leq M_Z^2$$

16.3 The Higgs sector of the standard model is highly arbitrary. Suppose that instead of a weak doublet the Higgs field is a weak triplet. Suppose further that the $T_3 = 0$ component develops a Vacuum Expectation Value. Find the hypercharge of the Higgs field that is necessary and following the treatment of eq.(16.22) show that $W^\circ$ and $\Gamma$ are unmixed and do not acquire mass from this VEV.

Suppose that the $T_3 = -1$ component develops a VEV. Show that in this case the treatment of (16.22) leads to the relation

$$M_{Z^\circ} = \frac{\sqrt{2}M_{W+}}{\cos\theta_W}$$

[This looks complicated but is very quick and easy.]

16.4* Calculate the total width of the $Z^\circ$, assuming that quarks can be treated as plane wave states. [Assume $u$, $d$, $s$, $c$, $b$ quarks and the three generations of leptons.] Suppose that further generations exist, the corresponding neutrinos being light relative to the $Z^\circ$. By how much does the $Z^\circ$ width increase for each additional generation of light neutrino?

[This looks as though it should be very easy but in fact contains some complicated features we have otherwise successfully circumvented. Suppose we work in the $Z^\circ$ centre of mass, with normalisation to 1/unit volume. The coupling to the fermion currents is given by

$$\frac{g}{\cos\theta_W}\bar{\psi}(c_V\gamma_\mu - c_A\gamma_\mu\gamma_5)\psi$$

but this must be contracted with a four-vector amplitude for destroying the $Z^\circ$. This amplitude is

$$\frac{\sqrt{4\pi}}{\sqrt{2E_Z}}\epsilon_\mu$$

where $\epsilon_\mu$ is a four vector describing the helicity state, normalised to unity. The factor $\frac{1}{\sqrt{2E_Z}}$ follows from the form of the probability density for the Klein-Gordon equation. The factor $\sqrt{4\pi}$ is a consequence of our choosing to define charges in terms of unrationalised units, so that the coulomb interaction, for example, is $e^2/r$ rather than $e^2/4\pi r$. In the rest frame, the four vectors $\epsilon_\mu$ reduce to three vectors. Helicity zero is (0,0,1), positive and negative helicity take the form $\frac{1}{\sqrt{2}}$ $(1, \pm i, 0)$. When all this is put in correctly and after integrating over the (obvious) angular terms, we obtain, for example,

$$\Gamma_{Z^\circ\to\nu\bar{\nu}} = \frac{4\pi e^2 M_Z}{96\pi\sin^2\theta_W\cos^2\theta_W}\left(= \frac{4\pi g^2}{24\pi}\frac{M_Z}{\cos^2\theta_W}\right)$$

(where $e^2 = 1/137$).]

16.5* Eq.(16.13) and Table 16.1 define the vector and axial vector couplings of the $Z^\circ$ to quarks and leptons. In $e^+e^-$ annihilation there is an amplitude corresponding to an intermediate $Z^\circ$ which is to be added to the amplitude corresponding to the intermediate photon (which has a pure vector coupling). Decompose $V$ and $A$ couplings into $L$ and $R$ couplings and hence construct the total amplitudes for

$$\begin{aligned} e_R^- e_L^+ &\to f_R \bar{f}_L && \text{(i)} \\ e_R^- e_L^+ &\to f_L \bar{f}_R && \text{(ii)} \\ e_L^- e_R^+ &\to f_L \bar{f}_R && \text{(iii)} \\ e_L^- e_R^+ &\to f_R \bar{f}_L && \text{(iv)} \end{aligned}$$

where $f_R$ is a right handed lepton or quark, $\bar{f}_R$ a right handed anti-fermion. From ch.9 you will recall that the angular factor associated with (i) and (iii) is $\cos^2\frac{\theta}{2}$; that associated with (ii) and (iv) is $\sin^2\frac{\theta}{2}$. The presence of $Z^\circ$ exchange unbalances the coefficients of $\cos^2\frac{\theta}{2}$ and $\sin^2\frac{\theta}{2}$ so that after squaring (i)–(iv) and summing the results, the differential cross section for (unpolarised) $e^+e^- \to f\bar{f}$ develops a term linear in $\cos\theta$. Obtain the following well known result

$$\begin{aligned} \frac{d\sigma}{d\Omega} &= \frac{e^4 f^2 N_c}{4s}\{(1+K_1)(1+\cos^2\theta) + K_2\cos\theta\} \\ K_1 &= 2c_V^e c_V^f \operatorname{Re}\chi + (c_V^{e^2} + c_A^{e^2})(c_V^{f^2} + c_A^{f^2})\chi^2 \\ K_2 &= 4c_A^e c_A^f \operatorname{Re}\chi + 8\chi^2 c_V^e c_A^e c_V^f c_A^f \end{aligned}$$

$$\chi = \frac{1}{f}\frac{1}{4\sin^2\theta_W\cos^2\theta_W}\frac{s}{s - M_Z^2 + iM_Z\Gamma_Z}$$

In the above expressions, $-f$ is the charge of the final fermion, in units of $e$, and $N_c$ is unity for final state leptons, three for final state quarks. At energies well below the $Z^\circ$ mass the value of $K_1$ is approximately zero and

$$K_2 \simeq 4c_A^e c_A^f \operatorname{Re}\chi$$

(since $c_V^e \simeq 0$).

Evaluate the forward-backward asymmetry

$$N_F - N_B/N_F + N_B$$

for final state leptons, charge $+\frac{2}{3}$ quarks and charge $-\frac{1}{3}$ quarks, at $\sqrt{s} =$ 29 GeV (PEP), 35 GeV (PETRA) and 54 GeV (TRISTAN). Consider how primary quark and antiquark directions might be distinguished for production of $c\bar{c}$ and $b\bar{b}$ pairs.

[You should find that the amplitude for (i) is modified from the intermediate photon case by the replacement

$$\frac{4\pi e^2 f}{s} \to \frac{4\pi e^2 f}{s} + \frac{4\pi g^2}{\cos^2\theta_W} \frac{c_R^e c_R^f}{s - M_Z^2 + iM_Z\Gamma_Z} \qquad \text{etc,}$$

where $c_R = c_V - c_A$.
The $Z°$ propagator may be taken as

$$\frac{(4\pi)}{s - M_Z^2 + iM_Z\Gamma_Z}$$

There are two subtleties here. First, the correct propagator for a massive spin 1 boson contains a term

$$\frac{k_\mu k_\nu}{M_Z^2}$$

in the numerator. This does not couple in the limit where the electron mass can be ignored, when

$$k_\nu \gamma_\nu \psi_e \approx 0.$$

Secondly, the width of the $Z°$ must be included for $s \sim M_Z^2$, yet it is a higher order correction. Simply inserting the width is no guarantee that these higher order corrections are handled correctly. Experimental measurements are now becoming sensitive to such radiative corrections.]

# FINAL REMARKS

This book has been concerned with the past. I have endeavoured to summarise the important features of a vast body of data and to present the physical notions which have given us a detailed understanding of so much. At the time of writing, I am aware of no phenomena which are clearly at variance with what is now called the standard model — $SU(3)_{colour} \times SU(2)_L \times U(1)_Y$, implemented with the minimal Higgs Sector. This is deeply frustrating, for this framework leaves undetermined a vast number of parameters and permits rather than requires a number of mysterious features. Prominent among these features are the existence of at least three generations, the parity violating $W$-fermion interactions (which are built into the model at the basement level) and the observed CP violation which has so far only been manifest in $K^\circ$ systems. The model does not relate the electroweak coupling constants $g$ and $g'$ of Ch.16 and neither is related to the colour couplings, yet all these interactions involve spin 1 bosons coupling to fermions. The successful prediction of the $Z^\circ$ mass relative to the charged $W$ mass is not as impressive as it seems, for the mysterious Higgs sector is highly arbitrary and at present unconstrained by experiment. The fermion masses are wholly arbitrary. There is no gravity in the standard model.

Nonetheless, the particle physics of today is the physics of fermions coupled by gauge bosons, fundamental at the level of $\lesssim 10^{-16}$cm. The problems of hadron physics are now regarded by most as merely computationally difficult and otherwise devoid of interest — spectroscopy, structure functions, fragmentation and the development of reggeon exchange in exclusive processes. The theoretical framework remains that delicate union of relativity and quantum mechanics known as quantum field theory. Quantum mechanics, devised originally in the context of atomic physics at a scale $\sim 10^{-9}$cm, has macroscopic manifestations — superconductivity and superfluidity, bulk ferromagnetism and not falling through the floor, among others — and applies equally well from macroscopic scales to the limit of resolution, $\sim 10^{-16}$cm. The heart of quantum mechanics is the addition of amplitudes — the principle of superposition — and we have encountered this core continually throughout the book. It works, yet quantum mechanics is not a realistic local theory and continues to defy our understanding.

The major theoretical development which has been driven by particle physics is the theory of self-interacting fields, covered by local gauge symmetries. The extent to which the idea of fields with vacuum expectation values is important must be left for the future to reveal. This idea has inspired the inflationary scenario for the very early universe, and one of the pleasing aspects of recent years is the growing relationship between particle physics and cosmology. The phenomenon of CP violation may be directly linked to the observations that matter in the universe consists of baryons (not antibaryons) and that the ratio of baryons to (black body) photons is $\sim 10^{-9}$. The problem of dark matter in the universe, apparently not baryonic matter, in turn may relate to unknown aspects of particle physics.

What of the future? The desired end is a theory of everything; a theory which embraces the standard model, yields relative coupling constants, masses and the number of generations, and embodies gravity. The first attempts at such a grand unification were less ambitious; to embed $\mathrm{SU}(3)_{colour}$, $\mathrm{SU}(2)_L$ and $\mathrm{U}(1)_Y$ in a larger group, thereby relating the three couplings of the standard model. Leptons and quarks then appear as members of the same multiplets and the exchange forces are enlarged. In addition to the gluons and electroweak bosons an additional class of bosons is necessary, carrying both leptonic and quark characteristics. They exchange quarks and leptons and in particular can mediate the conversion of quarks into leptons. In any such theory there is good reason to expect the proton to be unstable. [It is here that a possible explanation for the baryon asymmetry of the universe is encountered. A class of additional gauge bosons $X$ couples to $qq$ and with CP violation it could be that the rates $X \to qq$ and $\bar{X} \to \bar{q}\bar{q}$ are different. As the universe cooled below a temperature of $M_X$ a baryon asymmetry would develop; following annihilation of baryon and antibaryon at a later epoch baryons would remain, accompanied by (perhaps) $\sim 10^9$ annihilation photons for each baryon.] The three couplings of the standard model are related to a single coupling constant via the Clebsch-Gordan coefficients of the unifying group, but since the symmetry of the group is evidently broken in some way at present energies, the phenomenological couplings do not obey the simple relationships. The original toy model of this kind embedded the standard model within the group SU(5), but had the possibly undesirable feature that the quarks and leptons were distributed between two multiplets, a $\bar{5}$ and a 10.

$$\begin{aligned}\bar{5} &= (\nu_e, e)_L + \bar{d}_{rL} + \bar{d}_{bL} + \bar{d}_{gL} \\ 10 &= \bar{e}_L + \bar{u}_{rL} + \bar{u}_{bL} + \bar{u}_{gL} + (u_r, d_r)_L + (u_b, d_b)_L + (u_g, d_g)_L\end{aligned}$$

Extrapolation of the phenomenological couplings (which vary with energy) suggested true unification at $\sim 10^{14}$ GeV. The SU(5) Clebschs yield $\sin^2\theta_W = 3/8$ at this unification point; regression to low energy yielded $\sin^2\theta_W \simeq 0.21$ for the relation between the two phenomenological low energy electroweak couplings. If the unification scale is identified with the masses of the new leptoquark bosons, then one would obviously expect the proton decay rate to be

$$\Gamma_p \sim \frac{e^4}{M_X^4} \sim 10^{-60}\ \mathrm{GeV}$$

(with all other scales $\sim 1$ GeV) and hence

$$\tau_p \sim 10^{31}\ \mathrm{years}$$

The best experimental limits have come from watching $\sim 1000$ tonnes of water for $\sim 1$ year [Irvine-Michigan-Brookhaven and Kamioka experiments] and the (proper) SU(5) predictions are now excluded. The partial lifetimes for most decay modes are (experimentally) greater than $\sim 5 \times 10^{32}$ years. The SU(5) model is also regarded as being in trouble with $\sin^2\theta_W$, the prediction being about 10% lower than the measured value. It seemed like a good idea at the time, but it should be remembered that the group in which the standard model is embedded is not unique and that these relatively unambitious models are plagued by appalling Higgs sectors, determine neither masses nor the number of generations, and do not include gravity.

Another form of attempted unification seems at first sight ludicrous. Supersymmetry is often described as putting fermions and bosons into the same multiplet, so that in some sense they are indistinguishable. The idea is less ludicrous if you ask only that the coupling constants for fermions and bosons be related (through Clebschs) and it then has attraction in that the signs of boson loop amplitudes and fermion loop amplitudes are opposite. There is the prospect of cancelling the infinite loop integrals which are conventionally renormalised away, provided every fermion has a boson partner of equal mass (in the supersymmetric limit). A local gauged supersymmetry automatically generates gravity.

String theories were originally devised as a model of hadrons. A true string is a one dimensional object with constant energy in unit length and represents a step away from treating the most elementary objects as mass points. [Hadrons are not like this, although the long range flux tubes have some stringy properties.] The joining and parting of strings gives rise to highly constrained forms of interaction and the short distance behaviour which, with points, gives rise to divergences is tamed because the string has structure and can be excited. Consistent field theories of strings have not, to my knowledge, been constructed in less than 10 dimensions.

An enormous theoretical effort is now concentrated on superstring theory. There were indications that a consistent and possibly finite theory could be constructed with a unique prescription for the internal symmetry group (in which the standard model and more is to be embedded) and a unique way of curling up into invisibility the unwanted 6 dimensions. This program seems to have run into the sand — for the moment. I understand none of these developments.

It is important that grand unified theories, in which couplings are related to a single coupling via Clebsch-Gordan coefficients, yield quantisation of charge and with leptons and quarks in the same multiplet can yield $q_p = (-)q_e$. Plausible models yield reasonable values of $\sin^2 \theta_W$, but otherwise there has been little contact between unified models and the real world. The theoretical practitioners will learn much about field theory and there is already a symbiosis between this area of theoretical physics and more exotic areas of mathematics. We must wish them well in their endeavours, and hope for the emergence of a sense of direction from experiment.

When this book is first published substantial quantities of new data should have emerged from the Fermilab 1.8 TeV $p\bar{p}$ collider. The first physics should be emerging from the $Z^\circ$ phase of the $e^+e^-$ collider LEP at CERN, and it is to be hoped that the current problems will have been engineered out of the Stanford Linear Collider*. The Fermilab collider probably offers the best chance for identifying the missing top quark. The $e^+e^-$ colliders operating at the $Z^\circ$ peak will establish accurately the mass and width of the $Z^\circ$ (thereby setting a limit on the number of generations of light standard neutrinos) and the flood

---

* The first SLC $Z^\circ$ was reported in April 1989.

of data should further test the standard model as measurements of electroweak couplings are made with increasing precision. In 1990 the $e^{\pm}p$ collider HERA will probe the structure of the nucleon to a hitherto unrealised scale and will be able to probe for a component of the weak interaction coupling right handed to fermions. Later the second phase of LEP is expected to achieve energies of up to 200 GeV in the centre of mass, and in particular produce $W^+W^-$ pairs through the electroweak interactions. For the more distant future, proton colliders in the many TeV region are proposed in both the USA and Europe, and design studies for $e^+e^-$ linear colliders operating at energies above 1 TeV suggest that such machines could be built without radically new technology. We want the Higgs of course: otherwise we do not know what we want, save that it provide a sense of direction now lacking. We do not even know where such a catalyst is likely to be found ...

# UNITS

1. I have chosen to employ, for electromagnetism, the system in which the electric charge $q$ is defined by the equation

$$U_{12} = \frac{q_1 q_2}{r_{12}} \tag{U.1}$$

where $U_{12}$ is the potential energy of the system of charges and $r_{12}$ their separation. If a centimeter-gram-second system of units is used, the charge on the electron is $4.8 \times 10^{-10}$esu. Maxwell's equations take the form

$$\nabla \cdot \mathbf{E} = 4\pi\rho \qquad \nabla \cdot \mathbf{B} = 0$$
$$\nabla \times \mathbf{E} = -\frac{1}{c}\frac{\partial \mathbf{B}}{\partial t} \qquad \nabla \times \mathbf{B} = \frac{4\pi}{c}\mathbf{J} + \frac{1}{c}\frac{\partial \mathbf{E}}{\partial t}$$

where the charge density $\rho$ and current density $\mathbf{J}$ include any polarisation or magnetisation effects. Gauss' Law in electrostatics takes the form

$$\int \mathbf{E} \cdot \mathbf{ds} = 4\pi Q$$

where charge $Q$ is enclosed within the surface of integration.

The convention

$$U_{12} = \frac{q_1 q_2}{4\pi\epsilon_0 r_{12}} \tag{U.2}$$

is seldom used when discussing phenomena on a scale $\lesssim$ 1fm, but the (rationalised) choice

$$U_{12} = \frac{q_1 q_2}{4\pi r_{12}} \tag{U.3}$$

is frequently used. If $e$ is the charge on the electron, then †

$$(\mathrm{U.1}) \rightarrow (e^2/\hbar c) = 1/137 \ ^{\dagger}$$
$$(\mathrm{U.2}) \rightarrow (e^2/4\pi\epsilon_0\hbar c) = 1/137$$
$$(\mathrm{U.3}) \rightarrow (e^2/4\pi\hbar c) = 1/137$$

2. It is extremely convenient to work with momentum in units of energy/c, mass in units of energy/c$^2$, when the relation between energy, momentum and mass

$$p^2c^2 + m^2c^4 = E^2 \qquad \text{becomes} \qquad p^2 + m^2 = E^2$$

This choice may be represented by setting $c = 1$ (one light second per second, or one GeV$^{-1}$ per GeV$^{-1}$ — see below).

---

† More precisely, 1/137.035 989 5(61).

It is also very convenient to work with frequency and wavenumber rather than energy and momentum

$$\omega = E/\hbar \qquad k = p/\hbar$$

and choosing units such that $\hbar = 1$ (one GeV–GeV$^{-1}$). With the choice $\hbar = c = 1$ all dimensional quantities have dimensions which are powers of, for example, an energy. In high energy physics the unit of energy is usually 1 GeV, the energy gained by one electron charge falling through a potential difference of $10^9$ Volts. With the choice $\hbar = c = 1$, which is introduced progressively in this book, the dimensions of mass, momentum and energy are all GeV, the dimensions of distance and time are GeV$^{-1}$. A cross section has dimensions GeV$^{-2}$ and decay rate GeV.

In these units

$$e^2 = 1/137.$$

## GLOSSARY

A brief explanation of some terms which require no explanation to the experienced but which may be unfamiliar to some.

| | |
|---|---|
| C | frequently denotes the operation of charge conjugation, replacing every particle by its anti-particle, or the eigenvalue of the operator. |
| CPT | denotes the successive operations of time reversal, parity and charge conjugation. Relativistic quantum field theory is invariant under this portmanteau operation. |
| CERN | [*from* Conseil Européen pour la Recherche Nucléaire,] European Laboratory for Particle Physics, Geneva, Switzerland. |
| DESY | Deutsches Elektronen-Synchrotron Hamburg, Federal Republic of Germany. |
| $\Gamma$ | usually denotes the full width at half maximum of a resonance or an unstable state. |
| $\hbar$ | Planck's constant divided by $2\pi$; $6.582122 \times 10^{-25}$ GeV s |
| HERA | Hadron-Electron Ring Accelerator, at DESY. |
| $J$ (or $j$) | Total angular momentum, also current density. |
| $J^{PC}$ | specifies total angular momentum, parity and charge conjugation eigenvalues. |
| $J_Z$ (or $J_3$) | denotes the well defined projection. |
| LEP | Large Electron-Positron Collider, at CERN. |
| $L$ (or $l$) | Orbital angular momentum. |
| $P$ | frequently denotes the parity operation (co-ordinate inversion) or the eigenvalue of the operator. |
| $P_\ell\,(X)$ | Legendre polynomial of order $\ell$ in $X$. |
| PETRA | Positron Elektron Tandem Ring Anlage, at DESY. |
| QCD | Quantum Chromodynamics. |
| QED | Quantum Electrodynamics. |
| QWD | sometimes denotes the weak interactions. |
| S | may denote spin angular momentum *or* Strangeness |
| s | may denote orbital angular momentum zero, in the spectroscopic notation *spdf*... for orbital angular momentum, 0, 1, 2, 3 ... in units of $\hbar$;<br>*or* spin angular momentum<br>*or* the square of centre of mass energy |
| SLAC | Stanford Linear Accelerator Center, Stanford, California, U.S.A. |
| SLC | Stanford Linear Collider, at SLAC. |
| T | denotes the operation of time reversal<br>*or* the isospin of a state. |
| $T_Z$ (or $T_3$) | denotes the value of the third component. |

$T^{C_n} J^P$ denotes isospin, charge conjugation number of the neutral member of the multiplet, total angular momentum and parity.

$\tau$ denotes the proper mean life of an unstable particle or state *and* is the symbol used for the third generation charged lepton.

## SELECT BIBLIOGRAPHY


### Books

**1. General**

Introduction to High Energy Physics D.H. Perkins
Addison-Wesley 3rd Edition (1987)

Elementary Particle Physics I.R. Kenyon
Routledge and Kegan Paul (1987)

**2. Primarily theoretical**

Gauge Theories in Particle Physics I.J.R. Aitchison and A.J.G. Hey
Adam Hilger 2nd Edition (1989)

Quarks and Leptons F. Halzen and A.D. Martin
Wiley (1984)

Relativistic Quantum Mechanics J.D. Bjorken and S.D. Drell
McGraw-Hill (1964)

Relativistic Quantum Fields J.D. Bjorken and S.D. Drell
McGraw-Hill (1965)

Quantum Field Theory C. Itzykson and J-B. Zuber
McGraw-Hill (1980)

**3. Specialist**

Experimental Techniques in High Energy Physics T. Ferbel (Ed.)
Addison-Wesley (1987)

Hadron Interactions P.D.B. Collins and A.D. Martin
Adam Hilger (1984)

Weak Interactions of Leptons and Quarks E.D. Commins and P.H. Bucksbaum
Cambridge University Press (1983)

Grand Unified Theories G.G. Ross
Benjamin/Cummings (1985)

### Annual

Annual Review of Nuclear and Particle Science

### Journals

(i) Review: Physics Reports, Reviews of Modern Physics

(ii) The principal journals which carry papers on current results in particle physics are:

Physical Review Letters, Physical Review D, Physics Letters B, Zeitschrift für Physik C—Particles and Fields.

## Compendium

Review of Particle Properties, prepared by the Particle Data Group, currently appears every other year. The most recent edition is *Physics Letters* **B204** (1988).

This invaluable publication contains evaluated data on the properties of all known particles — and some unknown — with complete references and frequently minor review articles. The section entitled Miscellaneous Tables, Figures and Formulae is a handbook for the high energy physicist.

# INDEX